AF594983

2D Materials and van der Waals Heterostructures

2D Materials and van der Waals Heterostructures: Physics and Applications

Special Issue Editor

Antonio Di Bartolomeo

MDPI • Basel • Beijing • Wuhan • Barcelona • Belgrade • Manchester • Tokyo • Cluj • Tianjin

Special Issue Editor
Antonio Di Bartolomeo
Università di Salerno
Italy

Editorial Office
MDPI
St. Alban-Anlage 66
4052 Basel, Switzerland

This is a reprint of articles from the Special Issue published online in the open access journal *Nanomaterials* (ISSN 2079-4991) (available at: https://www.mdpi.com/journal/nanomaterials/special_issues/2d_materials_heterostructures).

For citation purposes, cite each article independently as indicated on the article page online and as indicated below:

LastName, A.A.; LastName, B.B.; LastName, C.C. Article Title. *Journal Name* **Year**, *Article Number*, Page Range.

ISBN 978-3-03928-768-0 (Hbk)
ISBN 978-3-03928-769-7 (PDF)

Contents

About the Special Issue Editor

Antonio Di Bartolomeo is professor of experimental condensed matter physics at Salerno University, Italy where he teaches semiconductor device physics, electric circuits, and electronics. His present research interests include: optical and electrical properties of nanostructured materials such as carbon nanotubes, graphene, and 2D materials; van der Waals heterojunctions of layered materials and Schottky heterojunctions; and field-effect transistors, non-volatile memories, solar cells, photodetectors, and field emission devices. He received his Ph.D. in physics in 1997 from Salerno University where he held the position of researcher in experimental physics for 16 years before the appointment as a professor. His scientific career started at CERN (CH) with the collaboration on experiments on neutrino oscillations and heavy ion collisions. He spent several years in the industry as a semiconductor device engineer (ST Microelectronics, Infineon Technologies, and Intel Corporation) and was guest scientist at IHP-Microelectronics (Germany) and Georgetown University (Washington, DC). He has authored more than 100 publications in peer-reviewed journals, two physics textbooks, and two patents, and has served as an Editorial Board member of several journals including *Nanomaterials*, by MDPI, *Nanotechnology* by IOP, and *Micro & Nano Letters* by IET.

Preface to "2D Materials and van der Waals Heterostructures: Physics and Applications"

The advent of graphene and two-dimensional (2D) layered materials has opened new perspectives in electronics, optoelectronics, energy harvesting, and sensing applications. 2D materials can be fabricated with relatively inexpensive production methods, integrated into existing semiconductor technologies, and offer new physical, chemical, and mechanical properties. Electrically, they can behave as insulators, semiconductors, metals, or even superconductors. Layered materials consist of covalently bonded and dangling-bond free monolayers that can be stacked on top of each other and held together by van der Waals forces. The number of layers can be controlled to tailor for specific properties. Different types of 2D materials can form heterojunctions with each other or with bulk materials, without the need for close lattice matching. In these heterojunctions, the weak van der Waals forces between the participant materials do not introduce significant changes at the atomic scale and essentially maintain the original electronic structure of the materials. Hence, van der Waals heterojunctions offer the opportunity to combine layers with different properties as the building blocks to engineer new functional materials for high-performance device or sensor applications. A notable advantage is that the easy stacking of a variety of 2D materials allows a far greater number of combinations than any traditional growth method. A tremendous amount of work has been conducted thus far on the physical and chemical properties as well as on the synthesis and the characterization of 2D materials such as graphene, transition metal dichalcogenides, hexagonal boron nitride, black phosphorus, silicene, organic perovskites, etc. Many of these materials have been used to fabricate stacked 2D–2D heterostructures, 2D/3D heterojunctions with common bulk semiconductors, or even 0D–2D and 1D–2D hybrids. The underlying physics and the possible applications in photodetection, biochemical sensing, strain gauges, photovoltaic energy generation, or energy harvesting have attracted the attention of both theorists and experimentalists. This book collects the papers published by Nanomaterials—MDPI as part of the Special Issue "2D Materials and Van der Waals Heterostructures: Physics and Applications". It covers state-of-the-art experimental, simulation, and theoretical research on 2D materials and their van der Waals heterojunctions with emphasis on the physical properties and the applications as sensors and electronic or optoelectronic devices.

Antonio Di Bartolomeo
Special Issue Editor

Editorial

Emerging 2D Materials and Their Van Der Waals Heterostructures

Antonio Di Bartolomeo

Physics Department "E.R.Caianiello" and "Interdepartmental center NANOMATES", University of Salerno, Fisciano, 84084 Salerno, Italy; adibartolomeo@unisa.it; Tel.: +39-089-96-9189

Received: 13 March 2020; Accepted: 17 March 2020; Published: 22 March 2020

Abstract: Two-dimensional (2D) materials and their van der Waals heterojunctions offer the opportunity to combine layers with different properties as the building blocks to engineer new functional materials for high-performance devices, sensors, and water-splitting photocatalysts. A tremendous amount of work has been done thus far to isolate or synthesize new 2D materials as well as to form new heterostructures and investigate their chemical and physical properties. This article collection covers state-of-the-art experimental, numerical, and theoretical research on 2D materials and on their van der Waals heterojunctions for applications in electronics, optoelectronics, and energy generation.

Keywords: graphene; MXene; transition metal dichalcogenide; van der Waals heterostructure; heterojunction; photodetection; photovoltaics; water splitting; photocatalysis

1. Introduction

The advent of graphene [1–3], and more recently of two-dimensional (2D) layered materials [4], has opened new perspectives in electronics, optoelectronics, energy generation, and sensing applications [5]. Two-dimensional materials can be fabricated with relatively inexpensive production methods, integrated into existing semiconductor technologies, and offer new physical and chemical properties [6–8]. Electrically, they can behave as insulators, semiconductors, metals, or even superconductors. Layered materials consist of covalently bonded and dangling-bond-free layers that can be stuck on top of each other by van der Waals forces to form bulk structures. In general, the number of layers can be controlled to tailor specific properties [9,10]. The possibility to accurately predict the physical properties of layered materials with the exact number of layers is a unique opportunity for directing the design and the fabrication of new electronic and optoelectronic devices.

Different types of 2D materials can form heterojunctions with each other or with bulk materials, without the need for close lattice matching [11]. In these heterojunctions, the weak van der Waals forces between the participant materials do not introduce significant changes at the atomic scale and usually maintain the original electronic structure of the materials. Hence, van der Waals heterojunctions offer the opportunity to combine layers with different properties as the building blocks to engineer new functional materials for high-performance electronic devices, chemical sensors, or water-splitting photocatalysts. A great advantage is that the easy stacking of a variety of 2D materials allows a far greater number of combinations than any traditional growth method [12].

Tremendous amount of work has been done thus far on the physical and chemical properties as well as on the synthesis and the characterization of 2D materials such as graphene [13,14], transition metal chalcogenides [15,16] and dichalcogenides [17–19], hexagonal boron nitride [20], black phosphorus [21], organic perovskites [22], etc. Many of these materials have been used to fabricate stacked 2D–2D heterostructures [23], 2D-3D heterojunctions with common bulk semiconductors [24,25], or even 0D–2D and 1D–2D hybrids [26]. The underlying physics and the possible applications in photodetection, biochemical sensing, strain gauges, photovoltaic energy generation, and photocatalytic water splitting have attracted the attention of both theorists and experimentalists.

This article collection, a reprint of the Special Issue "2D Materials and Van der Waals Heterostructures: Physics and Applications" published by *Nanomaterials* (MDPI), covers state-of-the-art experimental, simulation, and theoretical research on 2D materials and on their van der Waals heterojunctions for applications in electronics, optoelectronics, energy generation, and photocatalysis.

2. Emerging 2D Materials and Their Heterostructures

From a material standpoint, the articles of this collection can be organized in three main categories: Graphene and graphene oxide, MXenes and transition metal (di)chalcogenides, and graphene-like materials.

2.1. Graphene and Graphene Oxide

First principle calculations are applied to study the electronic and magnetic properties of Stone–Wales defected graphene [27] and the optical properties of graphene/MoS_2 heterostructures [28], while experimental work is carried out to investigate the properties of graphene/Si Schottky junctions [29] and to realize visible-light driven photoanodes for water oxidation [30].

Stone–Wales, formed by the rotation of a C-C bond in a hexagon ring, are the most common topological defects in graphene. They have been widely studied because they can lead to the opening of a band gap that is highly desirable for the application of graphene in electronic devices [31]. Stone-Wales defected graphene (SWG) includes two pairs of pentagonal–heptagonal rings and absorbs molecules more easily than perfect graphene, a feature important for sensor applications. The paper by Xie and coworkers [27] shows that it is possible to tune the band structure of SWG through the interaction with cyclopentadienyl and half-metallocene of Fe, Co, or Ni. The introduction of cyclopentadienyl and half-metallocene increases the conductivity SWG and induces magnetic properties, contributed by the 3d orbital of Fe, Co, Ni, and the molecular orbital of cyclopentadienyl. The study shows that the density of states and the magnetic properties in SWG can be tuned by controlling the cyclopentadienyl and half-metallocene absorption sites.

First principle calculations are also applied by Qiu et al. [28] to demonstrate that the electronic structure and the optical properties of graphene and monolayer molybdenum disulfide (MoS_2) are changed after they are combined in an heterostructure. MoS_2 [32,33] is the best known material of the transition metal dichalcogenide (TMD) family that will be treated next. Qiu et al. show that the optical properties of the graphene/MoS_2 system are improved compared to those of the two separate single-layers. The band gap and the dielectric constants become larger for the graphene/MoS_2 heterostructure, with redshift for the absorption coefficient, the refractive index, and the reflectance, and blueshift for the energy loss spectrum.

The heterojunction formed by graphene with traditional 3D materials, which has been a promising research topic [34–37], is experimentally investigated by Luongo and coworkers [35]. They fabricate graphene/n-Si junctions by transferring graphene on the flat surfaces of Si nanopillars, etched into a Si substrate, and obtain devices with rectifying behavior, remarkable photo-response, and photovoltaic capability. As is typical of good quality graphene/Si junctions [38], their devices exhibit a strongly bias- and temperature-dependent reverse current. Indeed, they report an exponentially growing reverse current below room temperature, which is explained as Schottky barrier lowering caused by the pillar-enhanced electric field, and a quasi-saturated reverse current at higher temperatures, attributed to the dominant effect of carrier thermal generation.

Graphene oxide is the material produced by oxidation of graphite that includes oxygen functional groups. It has interesting properties that are different than those of graphene. The reduction of the oxygen content through chemical, thermal, and other methods leads to the so-called reduced graphene oxide (r-GO), a cheaper and lower quality form of graphene, yet with important applications [39,40]. r-GO is used by Shuang et al. [30] for the fabrication of photoanodes for water splitting. Sustainable and convenient methods to prepare hydrogen fuel are a worldwide goal. Water splitting through the exploitation of solar energy is the most appealing method to obtain hydrogen fuel. The standard photocatalytic process for water splitting can be facilitated by the application of an external potential in the so-called photoelectrochemical approach. For this reason, many research efforts have been addressed to the fabrication of semiconducting nanomaterials with enhanced photoelectrochemical properties under visible light and, in particular, to the development of materials for photoanodes. Shuang et al. [30] propose a novel composite material consisting of β-$Cu_2V_2O_7$ nanoparticles deposited on TiO_2 nanorods followed by the addition of r-GO flakes. They show that electrophoretic deposition of p-type r-GO flakes on β-$Cu_2V_2O_7/TiO_2$ nanorods remarkably improves the durability, charge transfer resistance, and photocurrent density.

2.2. MXenes and Transition Metal (di)chalcogenides

MXenes are a new, large family of layered materials consisting of transition metal carbides, nitrides, and carbonitrides. Examples of mono and double transition metal MXenes are Ti_2C, Ti_4N_3, Ti_3CN, Mo_2N, Mo_2TiC_2, $Mo_2Ti_2C_3$, Cr_2TiC_2, etc. MXenes such as Nb_2CT_x, Ti_2CT_x, or $Ti_3C_2T_x$, with the surface terminated by T_x functional groups (e.g., O, F, OH, Cl), combine a high metallic conductivity with the hydrophilic nature of their hydroxyl or oxygen terminated surfaces, behaving as a sort of "conductive clays". MXenes have been applied to energy storage, water purification, chemical catalysts, electrochemical sensors, field effect transistor sensors, gas sensors, etc. [41].

Transition metal (mono)chalcogenides (TMC) [42] are MS, MSe, and MTe compounds (sulfides, selenides, tellurides of transition metals, M, such as titanium, niobium, molybdenum, cadmium, tungsten, etc.) and differ considerably from the transition metal oxides, MO, both in their structures and chemical and physical properties. Compared to oxygen atoms, the S, Se, and Te chalcogen atoms are larger, less electronegative, and have d orbitals (3d for S, 4d for Se, 5d for Te). Consequently, the metal-chalcogen bonds are more covalent than the metal-oxygen bonds and often involve the d orbitals. The increased covalency leads to broad valence and conduction bands with an energy gap generally narrower than in the case of oxides. Typically, the gap in sulfides is 1–3 eV, becomes smaller in selenides, and can vanish in tellurides. The availability of d orbitals in the calchogens causes large polarizability of chalcogenide ions. Furthermore, the extended d orbitals of the chalcogens can mix with metal d orbitals and have a stabilizing effect, causing for instance the smaller solubility of MS, MSe, and MTe compared to MO products in water.

Transition metal dichalcogenides (TMDs) are the most studied 2D materials after graphene [17,43]. They consist of a monolayer of transition metal atoms sandwiched between two layers of chalcogen atoms in a hexagonal or pentagonal lattice. Their standard structural formula is MX_2, where M represents a transition metal and X denotes the chalcogen. Molybdenum disulfide (MoS_2) and diselenide ($MoSe_2$) [44–48] and tungsten disulfide (WS_2) and diselenide (WSe_2) [49,50] are the most common TMDs with a hexagonal structure, palladium diselenide ($PdSe_2$) [10,51,52] is the prototype of TMDs with a pentagonal structure. Peculiarities such as absence of a dangling bond, electrocatalytic properties, chemical stability, mechanical flexibility, strong coupling to light, and bandgap tunable by number of layers and ranging from semi-metallic to over 2 eV, make TMDs materials of choice for the development of new sensors, electronic devices, and water-splitting photocatalysts. Indeed, different types of TMD-based biological sensors [53], flexible gas sensors [54], 3d image sensors [55], and tactile sensors [56], as well as large-scale electronic devices and circuits including logic, memory, optoelectronic, and analog devices [57] or photocatalysts for use in pollutant degradation and hydrogen evolution [58], have been demonstrated.

Xu et al. [59] propose $Ti_3C_2T_x$ MXene in combination with transition metal dichalcogenides to design new optical sensors based on surface plasmon resonance (SPR) to use for biosensing and chemical sensing. Highly sensitive SPR sensors have been previously demonstrated with coating of dielectric materials, graphene, or TMDs on metal films [59,60]. The new prism-coupled SPR sensor is based on Au-$Ti_3C_2T_x$-Au-TMDs in a modified Kretschmann configuration. The theoretical works of the authors show that it possesses enhanced sensitivity as compared to the bare Au film-based SPR sensor.

The synthesis of large area and good quality TMDs is of paramount importance for their industrial exploitation. The fabrication of monolayer WS_2 flakes is the subject of the work of Shi and coworkers [61]. WS_2 has a structure similar to the best-known MoS_2, but exhibits stronger photoluminescence quantum yield at room temperature and larger spin-orbit coupling and is a promising 2D material for applications in optoelectronics and spintronics. The difficulty of controlling the interrelated growth parameters makes the synthesis of large-area monolayer WS_2 still challenging. Therefore, the reported one-step chemical vapor deposition (CVD) process, by direct sulfurization of powdered tungsten trioxide (WO_3) drop-casted on SiO_2/Si substrates at atmospheric pressure that yields large-area monolayer WS_2, is of great interest. Triangular monolayer WS_2 flakes with edge lengths up few hundred microns and homogeneous crystallinity are obtained.

CVD monolayer WSe_2, which is a TMD structurally similar to MoS_2, but with a slightly lower bandgap (1.6 eV), is used in back-gated field effect transistors (FETs) by Urban et al. [62]. Two-dimensional WSe_2 exhibits strong optical absorption in the visible range, a good light-to-electricity conversion coefficient, and forms FETs with ambipolar behavior with most of the metals commonly used in electronics. Urban and coworkers fabricate WSe_2 backgate FETs with Ni Schottky contacts and measure their electrical characteristics under different environmental conditions. They demonstrate that lowering the pressure in air-exposed WSe_2 dramatically affects the electrical characteristics turning the FET conduction from the p- to n-type. Furthermore, from the electrical characterization at different temperatures, a gate modulation of the Schottky barrier (SB) at the contacts was proven. In addition, the work reports the temperature dependence of the carrier mobility and the subthreshold swing, as well as the photo response at several laser wavelengths. Some of the results are likely qualitatively valid for other TMD-based devices.

First principle calculations are again performed to investigate the electronic and optical properties of the β-polytypes of indium selenide, β-InSe [63], and carbon selenide, β-CSe [64], as well as to study the properties of Pd_2Se_3 [65]. Recently, much attention has been placed on the first two materials which belong to the family of layered TMC [66].

β-InSe is the most stable phase of InSe, due to the ABAB crystal stacking mode. The individual layer has a hexagonal structure and shows different electronic and optical properties compared to the bulk [40]. Monolayer and few-layer β-InSe possess moderate band gaps of 2.4 eV and 1.4 eV, respectively, and can be optimal candidates for use in broadband optoelectronic devices. Moreover, the appreciable shift of the valence band maximum upon thickness variation is very important for optimizing the band gap and improving the mobility of electrons and holes. The study presented by Sang et al. [63] shows that the electronic band structure, the work function, and optical properties of β-InSe are strongly dependent on the number of layers. For instance, the band structure exhibits direct-to-indirect transition from bulk β-InSe to few-layer β-InSe and the work functions varies in the 4.77–5.22 eV range depending on the number of layers. Similarly, the thickness variation strongly affects the imaginary part of the dielectric function which determines the optical properties of the material.

Based on numerical work, Zhang at al. propose carbon selenide, β-CSe, as a novel ultra-thin stable material with wide indirect bandgap. The β-CSe exhibits slightly anisotropic mechanical characteristics and its bandgap and band-edge curvature are sensitive to in-plane strain, causing the carrier effective mass to be strain-dependent. Zhang and coworkers also propose a heterojunction obtained by stacking α-CSe on a β-CSe sheet. The α-CSe/β-CSe interface constitutes a type-II van der Waals p-n heterojunction with strong built-in electric field across the interface, due to the charges transferring from β-CSe to α-CSe. The built-in potential causes substantial energy band bending, which can be exploited to spatially separate photo-generated carriers in photovoltaic devices. Furthermore, as a metal-free photocatalyst, the α-CSe/β-CSe heterojunction is endowed with an enhanced solar-driven redox ability for photocatalytic water splitting via reduced electron-hole-pair recombination. This study will likely motivate experimental research on the synthesis and the properties of this new material as well as on the design of new devices.

Layered materials based on noble transition metals, such as Pd and Pt, have gained popularity after the successful exfoliation of $PdSe_2$ [51,67,68] and the discovery of its stability in air. Following the observation of the formation of Pd_2Se_3 by interlayer fusion of $PdSe_2$ [69,70], Li and coworkers [65] study the physical properties and device applications of monolayer Pd_2Se_3 by first principles calculations. They demonstrate that Pd_2Se_3 monolayers have a great potential as absorber material in ultrathin photovoltaic devices owing to their quasi-direct band gap of 1.39 eV, high electron mobility (> 100 $cm^2V^{-1}s^{-1}$), and strong optical absorption (~10^5 cm^{-1}) in the visible solar spectrum. Furthermore, the Pd_2Se_3 bandgap can be modulated and changed from indirect to direct by biaxial strain. More importantly, monolayer Pd_2S_3, obtained by replacing Se with S, is stable and can be used in a vertical stack with Pd_2Se_3 monolayer to form a type-II heterostructure. The simulation indicates that such a structure, used as a solar cell, could achieve 20% power conversion efficiency, higher than that of $MoS_2/MoSe_2$ photovoltaic devices. This numerical study presents several challenges for new experiments on 2D Pd_2Se_3 and Pd_2S_3 as well as on the application of their heterostructures for efficient solar energy conversion.

2.3. Graphene-like Materials

The quest for efficient water-splitting photocatalysts motivates the first principle studies by Wang and coworkers [71,72], in which graphene-like carbon nitride, g-C_3N_4, and zinc oxide, g-ZnO, are used to form 2D heterojunctions with transition metal (di)chalcogenides. The investigation of CdS/g-C_3N_4 [71] and g-ZnO/WS_2 [72] heterostructures aims to find appropriate strategies to modulate the electronic and photocatalytic properties of the individual materials. Indeed, both cadmium sulfide CdS [73] and graphitic carbon nitride g-C_3N_4 [74] have suitable bandgaps for visible light absorption (2.4 and 2.7 eV, respectively) and could be good photocatalysts, but the former lacks stability due to the self-oxidation of photogenerated species while the latter has poor photocatalytic efficiency because of the fast recombination of photogenerated electron–hole pairs. However, significantly improved photocatalytic activity, as compared to the individual 2D materials, can be achieved with their heterostructure. The built-in electric field, due to the charge accumulation/depletion around the interfaces, promotes the effective separation and migration of photogenerated carriers, which is beneficial for the photocatalytic performance. Hence, Wang et al. investigate the energetic, electronic and optical properties, and the band edge alignments of CdS/g-C_3N_4 as a function of biaxial strain and pH of the electrolyte, with the objective of optimizing the photocatalytic performances. The numerical simulations show that monolayer CdS weakly contacts with g-C_3N_4, forming a type II van der Waals (vdW) heterostructure. The predicted bandgaps and optical absorptions indicate that such a heterostructure can absorb visible light and the induced built-in electric field at the interface promotes the effective separation of the photogenerated carriers. Furthermore, the interface adhesion energy, bandgap, and band edge positions can be adjusted by applying a biaxial strain.

The water-splitting photocatalytic properties of the heterostructures formed by g-ZnO with two-dimensional WS_2 or WSe_2 are studied as a function of the rotation angles and the biaxial strain [72] as well. g-ZnO has been experimentally synthesized [75] and density functional theory studies have suggested that doping with nonmetal species can endow it with tunable magnetism [76]. Wang and coworkers demonstrate that g-ZnO/WS_2 heterostructures with appropriate rotation angles possess a suitable bandgap for visible absorption, proper band edge alignment, and effective separation of carriers, being promising visible water-splitting photocatalysts. Conversely, the water oxygen process of the ZnO/WSe_2 heterostructures is limited by their band edge positions.

These theoretical findings offer a sound basis to develop CdS/g-C_3N_4 or g-ZnO/WS_2-based water-splitting solutions to prepare hydrogen fuel.

3. Conclusions

Although graphene has proven to be an unmatched material due to its abundance of properties and applications, other 2D materials have been isolated or synthesized to better serve specific needs. Owing to their bandgaps, an example of a feature that graphene does not possess, transition metal (di)chalcogenides have emerged for instance as materials well-suited for electronics, optoelectronic, and photocatalytic applications.

The layer control in two-dimensional materials provides an effective strategy to modulate their optical and electrical properties. The formation of van der Waals heterostructures, with enhanced features, offers a unique opportunity to fabricate new materials.

This book offers a collection of research papers that cover the state-of-the-art of fabrication, properties, and applications of graphene and graphene-like materials, along with their heterostructures.

The electronic and optical properties of several popular 2D materials and their heterojunctions, as well as the opportunities for their exploitation in sensors, electronic and optoelectronic devices, and water-splitting photocatalysts, are extensively discussed. New 2D materials and new heterostructures like α-CSe/β-CSe, Pd_2Se_3/Pd_2S_3 or CdS/g-C_3N_4, and g-ZnO/WS_2 are proposed based on first principle calculations.

These studies consolidate and extend the general knowledge of the physics and technology of 2D materials and offer many theoretical results that will likely be the challenging subject of forthcoming experimental investigations.

Funding: This research was funded by MIUR, Project Pico & Pro, ARS01_01061, 2018-2021 and Project RINASCIMENTO, ARS01_01088, 2018-2021.

Acknowledgments: A special thank you to all the colleagues who submitted their research work to the Special Issue "2D Materials and Van der Waals Heterostructures: Physics and Applications".

Conflicts of Interest: The author declares no conflict of interest.

References

1. Novoselov, K.S. Electric Field Effect in Atomically Thin Carbon Films. *Science* **2004**, *306*, 666–669. [CrossRef] [PubMed]
2. Geim, A.K.; Novoselov, K.S. The rise of graphene. *Nat. Mater.* **2007**, *6*, 183–191. [CrossRef] [PubMed]
3. Bartolomeo, A.D.; Giubileo, F.; Santandrea, S.; Romeo, F.; Citro, R.; Schroeder, T.; Lupina, G. Charge transfer and partial pinning at the contacts as the origin of a double dip in the transfer characteristics of graphene-based field-effect transistors. *Nanotechnology* **2011**, *22*, 275702. [CrossRef]
4. Novoselov, K.S.; Jiang, D.; Schedin, F.; Booth, T.J.; Khotkevich, V.V.; Morozov, S.V.; Geim, A.K. Two-dimensional atomic crystals. *Proc. Natl. Acad. Sci. USA* **2005**, *102*, 10451–10453. [CrossRef] [PubMed]
5. Lin, Z.; McCreary, A.; Briggs, N.; Subramanian, S.; Zhang, K.; Sun, Y.; Li, X.; Borys, N.J.; Yuan, H.; Fullerton-Shirey, S.K.; et al. 2D materials advances: From large scale synthesis and controlled heterostructures to improved characterization techniques, defects and applications. *2D Materrials* **2016**, 3, 042001. [CrossRef]
6. Giannazzo, F.; Avila, S.; Eriksson, J.; Sonde, S. *Integration of 2D Materials for Electronics Applications*; MDPI: Basel, Switzerland, 2019; ISBN 978-3-03897-607-3.
7. Liu, L.; Zhou, M.; Li, X.; Jin, L.; Su, G.; Mo, Y.; Li, L.; Zhu, H.; Tian, Y. Research Progress in Application of 2D Materials in Liquid-Phase Lubrication System. *Materials* **2018**, *11*, 1314. [CrossRef]
8. Grillo, A.; Di Bartolomeo, A.; Urban, F.; Passacantando, M.; Caridad, J.M.; Sun, J.; Camilli, L. Observation of 2D Conduction in Ultrathin Germanium Arsenide Field-Effect Transistors. *ACS Appl. Mater. Interfaces* **2020**, *12*, 12998–13004. [CrossRef]
9. Li, X.-L.; Han, W.-P.; Wu, J.-B.; Qiao, X.-F.; Zhang, J.; Tan, P.-H. Layer-Number Dependent Optical Properties of 2D Materials and Their Application for Thickness Determination. *Adv. Funct. Mater.* **2017**, *27*, 1604468. [CrossRef]

10. Di Bartolomeo, A.; Pelella, A.; Liu, X.; Miao, F.; Passacantando, M.; Giubileo, F.; Grillo, A.; Iemmo, L.; Urban, F.; Liang, S. Pressure-Tunable Ambipolar Conduction and Hysteresis in Thin Palladium Diselenide Field Effect Transistors. *Adv. Funct. Mater.* **2019**, *29*, 1902483. [CrossRef]
11. Geim, A.K.; Grigorieva, I.V. Van der Waals heterostructures. *Nature* **2013**, *499*, 419–425. [CrossRef]
12. Hu, W.; Yang, J. Two-dimensional van der Waals heterojunctions for functional materials and devices. *J. Mater. Chem. C* **2017**, *5*, 12289–12297. [CrossRef]
13. Randviir, E.P.; Brownson, D.A.C.; Banks, C.E. A decade of graphene research: Production, applications and outlook. *Mater. Today* **2014**, *17*, 426–432. [CrossRef]
14. Giubileo, F.; Di Bartolomeo, A. The role of contact resistance in graphene field-effect devices. *Prog. Surf. Sci.* **2017**, *92*, 143–175. [CrossRef]
15. Shanmugaratnam, S.; Rasalingam, S. Transition Metal Chalcogenide (TMC) Nanocomposites for Environmental Remediation Application over Extended Solar Irradiation. In *Nanocatalysts*; Sinha, I., Shukla, M., Eds.; IntechOpen: London, UK, 2019; ISBN 978-1-78984-159-6.
16. Zhou, X.; Rodriguez, E.E. Tetrahedral Transition Metal Chalcogenides as Functional Inorganic Materials. *Chem. Mater.* **2017**, *29*, 5737–5752. [CrossRef]
17. Lv, R.; Robinson, J.A.; Schaak, R.E.; Sun, D.; Sun, Y.; Mallouk, T.E.; Terrones, M. Transition Metal Dichalcogenides and Beyond: Synthesis, Properties, and Applications of Single- and Few-Layer Nanosheets. *Acc. Chem. Res.* **2015**, *48*, 56–64. [CrossRef] [PubMed]
18. Urban, F.; Giubileo, F.; Grillo, A.; Iemmo, L.; Luongo, G.; Passacantando, M.; Foller, T.; Madauß, L.; Pollmann, E.; Geller, M.P.; et al. Gas dependent hysteresis in MoS_2 field effect transistors. *2D Materials* **2019**, *6*, 045049. [CrossRef]
19. Heine, T. Transition Metal Chalcogenides: Ultrathin Inorganic Materials with Tunable Electronic Properties. *Acc. Chem. Res.* **2015**, *48*, 65–72. [CrossRef]
20. Bhimanapati, G.R.; Glavin, N.R.; Robinson, J.A. 2D Boron Nitride. In *Semiconductors and Semimetals*; Elsevier: Amsterdam, The Netherlands, 2016; Volume 95, pp. 101–147. ISBN 978-0-12-804272-4.
21. Xu, Y.; Shi, Z.; Shi, X.; Zhang, K.; Zhang, H. Recent progress in black phosphorus and black-phosphorus-analogue materials: Properties, synthesis and applications. *Nanoscale* **2019**, *11*, 14491–14527. [CrossRef]
22. Li, L.; He, Y.; Xu, L.; Wang, H. Synthesis, Structure and Photoluminescence Properties of 2D Organic–Inorganic Hybrid Perovskites. *Appl. Sci.* **2019**, *9*, 5211. [CrossRef]
23. Novoselov, K.S.; Mishchenko, A.; Carvalho, A.; Castro Neto, A.H. 2D materials and van der Waals heterostructures. *Science* **2016**, *353*, aac9439. [CrossRef]
24. Di Bartolomeo, A. Graphene Schottky diodes: An experimental review of the rectifying graphene/semiconductor heterojunction. *Phys. Rep.* **2016**, *606*, 1–58. [CrossRef]
25. Di Bartolomeo, A.; Luongo, G.; Iemmo, L.; Urban, F.; Giubileo, F. Graphene–Silicon Schottky Diodes for Photodetection. *IEEE Trans. Nanotechnol.* **2018**, *17*, 1133–1137. [CrossRef]
26. Chen, J.-S.; Doane, T.L.; Li, M.; Zang, H.; Maye, M.M.; Cotlet, M. 0D-2D and 1D-2D Semiconductor Hybrids Composed of All Inorganic Perovskite Nanocrystals and Single-Layer Graphene with Improved Light Harvesting. *Part. Part. Syst. Charact.* **2018**, *35*, 1700310. [CrossRef]
27. Xie, K.; Jia, Q.; Zhang, X.; Fu, L.; Zhao, G. Electronic and Magnetic Properties of Stone–Wales Defected Graphene Decorated with the Half-Metallocene of M (M = Fe, Co, Ni): A First Principle Study. *Nanomaterials* **2018**, *8*, 552. [CrossRef] [PubMed]
28. Qiu, B.; Zhao, X.; Hu, G.; Yue, W.; Ren, J.; Yuan, X. Optical Properties of Graphene/MoS_2 Heterostructure: First Principles Calculations. *Nanomaterials* **2018**, *8*, 962. [CrossRef] [PubMed]
29. Luongo, G.; Grillo, A.; Giubileo, F.; Iemmo, L.; Lukosius, M.; Alvarado Chavarin, C.; Wenger, C.; Di Bartolomeo, A. Graphene Schottky Junction on Pillar Patterned Silicon Substrate. *Nanomaterials* **2019**, *9*, 659. [CrossRef] [PubMed]
30. Shuang, S.; Girardi, L.; Rizzi, G.; Sartorel, A.; Marega, C.; Zhang, Z.; Granozzi, G. Visible Light Driven Photoanodes for Water Oxidation Based on Novel r-GO/β-Cu2V2O7/TiO2 Nanorods Composites. *Nanomaterials* **2018**, *8*, 544. [CrossRef]
31. Bartolomeo, A.D.; Giubileo, F.; Romeo, F.; Sabatino, P.; Carapella, G.; Iemmo, L.; Schroeder, T.; Lupina, G. Graphene field effect transistors with niobium contacts and asymmetric transfer characteristics. *Nanotechnology* **2015**, *26*, 475202. [CrossRef]

32. Iemmo, L.; Urban, F.; Giubileo, F.; Passacantando, M.; Di Bartolomeo, A. Nanotip Contacts for Electric Transport and Field Emission Characterization of Ultrathin MoS_2 Flakes. *Nanomaterials* **2020**, *10*, 106. [CrossRef]
33. Krishnan, U.; Kaur, M.; Singh, K.; Kumar, M.; Kumar, A. A synoptic review of MoS_2: Synthesis to applications. *Superlattices Microstruct.* **2019**, *128*, 274–297. [CrossRef]
34. Di Bartolomeo, A.; Luongo, G.; Giubileo, F.; Funicello, N.; Niu, G.; Schroeder, T.; Lisker, M.; Lupina, G. Hybrid graphene/silicon Schottky photodiode with intrinsic gating effect. *2D Materials* **2017**, *4*, 025075. [CrossRef]
35. Luongo, G.; Giubileo, F.; Genovese, L.; Iemmo, L.; Martucciello, N.; Di Bartolomeo, A. I-V and C-V Characterization of a High-Responsivity Graphene/Silicon Photodiode with Embedded MOS Capacitor. *Nanomaterials* **2017**, *7*, 158. [CrossRef] [PubMed]
36. Di Bartolomeo, A.; Giubileo, F.; Luongo, G.; Iemmo, L.; Martucciello, N.; Niu, G.; Fraschke, M.; Skibitzki, O.; Schroeder, T.; Lupina, G. Tunable Schottky barrier and high responsivity in graphene/Si-nanotip optoelectronic device. *2D Materials* **2016**, *4*, 015024. [CrossRef]
37. Alvarado Chavarin, C.; Strobel, C.; Kitzmann, J.; Di Bartolomeo, A.; Lukosius, M.; Albert, M.; Bartha, J.; Wenger, C. Current Modulation of a Heterojunction Structure by an Ultra-Thin Graphene Base Electrode. *Materials* **2018**, *11*, 345. [CrossRef]
38. Luongo, G.; Di Bartolomeo, A.; Giubileo, F.; Chavarin, C.A.; Wenger, C. Electronic properties of graphene/p-silicon Schottky junction. *J. Phys. D Appl. Phys.* **2018**, *51*, 255305. [CrossRef]
39. Pei, S.; Cheng, H.-M. The reduction of graphene oxide. *Carbon* **2012**, *50*, 3210–3228. [CrossRef]
40. Smith, A.T.; LaChance, A.M.; Zeng, S.; Liu, B.; Sun, L. Synthesis, properties, and applications of graphene oxide/reduced graphene oxide and their nanocomposites. *Nano Mater. Sci.* **2019**, *1*, 31–47. [CrossRef]
41. Naguib, M.; Mochalin, V.N.; Barsoum, M.W.; Gogotsi, Y. 25th Anniversary Article: MXenes: A New Family of Two-Dimensional Materials. *Adv. Mater.* **2014**, *26*, 992–1005. [CrossRef]
42. Jellinek, F. Transition metal chalcogenides. relationship between chemical composition, crystal structure and physical properties. *React. Solids* **1988**, *5*, 323–339. [CrossRef]
43. Ravindra, N.M.; Tang, W.; Rassay, S. Transition Metal Dichalcogenides Properties and Applications. In *Semiconductors*; Pech-Canul, M.I., Ravindra, N.M., Eds.; Springer International Publishing: Cham, Switzerland, 2019; pp. 333–396. ISBN 978-3-030-02169-6.
44. Di Bartolomeo, A.; Genovese, L.; Giubileo, F.; Iemmo, L.; Luongo, G.; Foller, T.; Schleberger, M. Hysteresis in the transfer characteristics of MoS_2 transistors. *2D Materials* **2017**, *5*, 015014. [CrossRef]
45. Kong, D.; Wang, H.; Cha, J.J.; Pasta, M.; Koski, K.J.; Yao, J.; Cui, Y. Synthesis of MoS_2 and $MoSe_2$ Films with Vertically Aligned Layers. *Nano Lett.* **2013**, *13*, 1341–1347. [CrossRef] [PubMed]
46. Urban, F.; Passacantando, M.; Giubileo, F.; Iemmo, L.; Di Bartolomeo, A. Transport and Field Emission Properties of MoS_2 Bilayers. *Nanomaterials* **2018**, *8*, 151. [CrossRef] [PubMed]
47. Di Bartolomeo, A.; Genovese, L.; Foller, T.; Giubileo, F.; Luongo, G.; Croin, L.; Liang, S.-J.; Ang, L.K.; Schleberger, M. Electrical transport and persistent photoconductivity in monolayer MoS_2 phototransistors. *Nanotechnology* **2017**, *28*, 214002. [CrossRef] [PubMed]
48. Giubileo, F.; Iemmo, L.; Passacantando, M.; Urban, F.; Luongo, G.; Sun, L.; Amato, G.; Enrico, E.; Di Bartolomeo, A. Effect of Electron Irradiation on the Transport and Field Emission Properties of Few-Layer MoS_2 Field-Effect Transistors. *J. Phys. Chem. C* **2019**, *123*, 1454–1461. [CrossRef]
49. Da Silva, A.C.H.; Caturello, N.A.M.S.; Besse, R.; Lima, M.P.; Da Silva, J.L.F. Edge, size, and shape effects on WS_2, WSe_2, and WTe_2 nanoflake stability: Design principles from an ab initio investigation. *Phys. Chem. Chem. Phys.* **2019**, *21*, 23076–23084. [CrossRef] [PubMed]
50. Di Bartolomeo, A.; Urban, F.; Passacantando, M.; McEvoy, N.; Peters, L.; Iemmo, L.; Luongo, G.; Romeo, F.; Giubileo, F. A WSe_2 vertical field emission transistor. *Nanoscale* **2019**, *11*, 1538–1548. [CrossRef] [PubMed]
51. Oyedele, A.D.; Yang, S.; Liang, L.; Puretzky, A.A.; Wang, K.; Zhang, J.; Yu, P.; Pudasaini, P.R.; Ghosh, A.W.; Liu, Z.; et al. $PdSe_2$: Pentagonal Two-Dimensional Layers with High Air Stability for Electronics. *J. Am. Chem. Soc.* **2017**, *139*, 14090–14097. [CrossRef]
52. Pi, L.; Li, L.; Liu, K.; Zhang, Q.; Li, H.; Zhai, T. Recent Progress on 2D Noble-Transition-Metal Dichalcogenides. *Adv. Funct. Mater.* **2019**, *29*, 1904932. [CrossRef]
53. Ponnusamy, R.; Rout, C.S. Transition Metal Dichalcogenides in Sensors. In *Two Dimensional Transition Metal Dichalcogenides*; Arul, N.S., Nithya, V.D., Eds.; Springer: Singapore, 2019; pp. 293–329, ISBN 9789811390449.

54. Kumar, R.; Goel, N.; Hojamberdiev, M.; Kumar, M. Transition metal dichalcogenides-based flexible gas sensors. *Sens. Actuators A Phys.* **2020**, *303*, 111875. [CrossRef]
55. Yang, C.-C.; Chiu, K.-C.; Chou, C.-T.; Liao, C.-N.; Chuang, M.-H.; Hsieh, T.-Y.; Huang, W.-H.; Shen, C.-H.; Shieh, J.-M.; Yeh, W.-K.; et al. Enabling monolithic 3D image sensor using large-area monolayer transition metal dichalcogenide and logic/memory hybrid $3D^+$ IC. In Proceedings of the 2016 IEEE Symposium on VLSI Technology, Honolulu, HI, USA, 14–16 June 2016; pp. 1–2.
56. Park, M.; Park, Y.J.; Chen, X.; Park, Y.-K.; Kim, M.-S.; Ahn, J.-H. MoS_2 -Based Tactile Sensor for Electronic Skin Applications. *Adv. Mater.* **2016**, *28*, 2556–2562. [CrossRef]
57. Tang, H.; Zhang, H.; Chen, X.; Wang, Y.; Zhang, X.; Cai, P.; Bao, W. Recent progress in devices and circuits based on wafer-scale transition metal dichalcogenides. *Sci. China Inf. Sci.* **2019**, *62*, 220401. [CrossRef]
58. Huang, T.; Zhang, M.; Yin, H.; Liu, X. Transition Metal Dichalcogenides in Photocatalysts. In *Two Dimensional Transition Metal Dichalcogenides*; Arul, N.S., Nithya, V.D., Eds.; Springer: Singapore, 2019; pp. 107–134, ISBN 9789811390449.
59. Xu, Y.; Hsieh, C.-Y.; Wu, L.; Ang, L.K. Two-dimensional transition metal dichalcogenides mediated long range surface plasmon resonance biosensors. *J. Phys. D Appl. Phys.* **2019**, *52*, 065101. [CrossRef]
60. Xu, Y.; Wu, L.; Ang, L.K. MoS_2-Based Highly Sensitive Near-Infrared Surface Plasmon Resonance Refractive Index Sensor. *IEEE J. Select. Topics Quantum Electron.* **2019**, *25*, 1–7.
61. Shi, B.; Zhou, D.; Fang, S.; Djebbi, K.; Feng, S.; Zhao, H.; Tlili, C.; Wang, D. Facile and Controllable Synthesis of Large-Area Monolayer WS_2 Flakes Based on WO3 Precursor Drop-Casted Substrates by Chemical Vapor Deposition. *Nanomaterials* **2019**, *9*, 578. [CrossRef] [PubMed]
62. Urban, F.; Martucciello, N.; Peters, L.; McEvoy, N.; Di Bartolomeo, A. Environmental Effects on the Electrical Characteristics of Back-Gated WSe2 Field-Effect Transistors. *Nanomaterials* **2018**, *8*, 901. [CrossRef]
63. Sang, D.K.; Wang, H.; Qiu, M.; Cao, R.; Guo, Z.; Zhao, J.; Li, Y.; Xiao, Q.; Fan, D.; Zhang, H. Two Dimensional β-InSe with Layer-Dependent Properties: Band Alignment, Work Function and Optical Properties. *Nanomaterials* **2019**, *9*, 82. [CrossRef]
64. Zhang, Q.; Feng, Y.; Chen, X.; Zhang, W.; Wu, L.; Wang, Y. Designing a Novel Monolayer β-CSe for High Performance Photovoltaic Device: An Isoelectronic Counterpart of Blue Phosphorene. *Nanomaterials* **2019**, *9*, 598. [CrossRef]
65. Li, X.; Zhang, S.; Guo, Y.; Wang, F.; Wang, Q. Physical Properties and Photovoltaic Application of Semiconducting Pd2Se3 Monolayer. *Nanomaterials* **2018**, *8*, 832. [CrossRef]
66. Huang, W.; Gan, L.; Li, H.; Ma, Y.; Zhai, T. 2D layered group IIIA metal chalcogenides: Synthesis, properties and applications in electronics and optoelectronics. *CrystEngComm* **2016**, *18*, 3968–3984. [CrossRef]
67. Di Bartolomeo, A.; Pelella, A.; Urban, F.; Grillo, A.; Iemmo, L.; Passacantando, M.; Liu, X.; Giubileo, F. Field emission in ultrathin PdSe2 back-gated transistors. *arXiv* **2020**, arXiv:2002.05454.
68. Giubileo, F.; Grillo, A.; Iemmo, L.; Luongo, G.; Urban, F.; Passacantando, M.; Di Bartolomeo, A. Environmental effects on transport properties of PdSe2 field effect transistors. *Mater. Today Proc.* **2020**, *20*, 50–53. [CrossRef]
69. Lin, J.; Zuluaga, S.; Yu, P.; Liu, Z.; Pantelides, S.T.; Suenaga, K. Novel Pd2Se3 Two-Dimensional Phase Driven by Interlayer Fusion in Layered PdSe2. *Phys. Rev. Lett.* **2017**, *119*, 016101. [CrossRef]
70. Di Bartolomeo, A.; Urban, F.; Pelella, A.; Grillo, A.; Passacantando, M.; Liu, X.; Giubileo, F. Electron irradiation on multilayer PdSe2 field effect transistors. *arXiv* **2020**, arXiv:2002.09785.
71. Wang, G.; Zhou, F.; Yuan, B.; Xiao, S.; Kuang, A.; Zhong, M.; Dang, S.; Long, X.; Zhang, W. Strain-Tunable Visible-Light-Responsive Photocatalytic Properties of Two-Dimensional CdS/g-C3N4: A Hybrid Density Functional Study. *Nanomaterials* **2019**, *9*, 244. [CrossRef] [PubMed]
72. Wang, G.; Li, D.; Sun, Q.; Dang, S.; Zhong, M.; Xiao, S.; Liu, G. Hybrid Density Functional Study on the Photocatalytic Properties of Two-dimensional g-ZnO Based Heterostructures. *Nanomaterials* **2018**, *8*, 374. [CrossRef]
73. Cheng, L.; Xiang, Q.; Liao, Y.; Zhang, H. CdS-Based photocatalysts. *Energy Environ. Sci.* **2018**, *11*, 1362–1391. [CrossRef]
74. Dong, G.; Zhang, Y.; Pan, Q.; Qiu, J. A fantastic graphitic carbon nitride (g-C3N4) material: Electronic structure, photocatalytic and photoelectronic properties. *J. Photochem. Photobiol. C Photochem. Rev.* **2014**, *20*, 33–50. [CrossRef]

75. Tusche, C.; Meyerheim, H.L.; Kirschner, J. Observation of Depolarized ZnO(0001) Monolayers: Formation of Unreconstructed Planar Sheets. *Phys. Rev. Lett.* **2007**, *99*, 026102. [CrossRef]
76. Guo, H.; Zhao, Y.; Lu, N.; Kan, E.; Zeng, X.C.; Wu, X.; Yang, J. Tunable Magnetism in a Nonmetal-Substituted ZnO Monolayer: A First-Principles Study. *J. Phys. Chem. C* **2012**, *116*, 11336–11342. [CrossRef]

Article

Electronic and Magnetic Properties of Stone–Wales Defected Graphene Decorated with the Half-Metallocene of *M* (*M* = Fe, Co, Ni): A First Principle Study

Kefeng Xie [1,2,*], Qiangqiang Jia [2], Xiangtai Zhang [2], Li Fu [3] and Guohu Zhao [1,*]

1 Provincial Key Laboratory of Gansu Higher Education for City Environmental Pollution Control, School of Chemistry and Chemical Engineering, Lanzhou City University, Lanzhou 730070, China
2 State Key Laboratory of Plateau Ecology and Agriculture, Qinghai University, Xining 810016, China; 2015990037@qhu.edu.cn (Q.J.); 2017990017@qhu.edu.cn (X.Z.)
3 Collage of Materials and Environmental Engineering, Hangzhou Dianzi University, Hangzhou 310018, China; fuli@hdu.edu.cn
* Correspondence: xiekf12@lzu.edu.cn (K.X.), zhaoguohu@lzcu.edu.cn (G.Z.); Tel.: +86-931-7601139 (K.X.)

Received: 27 June 2018; Accepted: 17 July 2018; Published: 20 July 2018

Abstract: The geometrical, electronic structure, and magnetic properties of the half-metallocene of *M* (*M* = Fe, Co, Ni) adsorbed on Stone–Wales defected graphene (SWG) were studied using the density functional theory (DFT), aiming to tune the band structure of SWG. The introduction of cyclopentadienyl (Cp) and half-metallocene strongly affected the band structure of SWG. The magnetic properties of the complex systems originated from the 3d orbitals of *M* (*M* = Fe, Co, Ni), the molecular orbital of Cp, and SWG. This phenomenon was different from that found in a previous study, which was due to metal ion-induced sandwich complexes. The results have potential applications in the design of electronic devices based on SWG.

Keywords: Stone–Wales defected graphene; half-metallocene; adsorption energy; density of states; and magnetic property

---

1. Introduction

Graphene, which is as a typical two-dimensional (2D) material, has aroused considerable attention because of its special properties and promising potential applications in electronic devices, nanocomposites, molecule sensors, transparent electrodes in light emitting diodes (LED), and photovoltaic devices [1–4]. The typical structural defects [5] in graphene are vacancies [6], impurities [7–10], Stone–Wales (SW) defects [11–14], and pentagonal–octagonal defects [15,16]. The various defects of graphene can alter its electronic and mechanical properties significantly [5,17,18]. The most common topological defect in graphene is SW defect, which is formed by an in-plane 90° rotation of a C-C bond in a hexagon ring with fixing the middle point of this bond. SW defects in graphene have been studied widely because the opening band gap of electronic structure is applied to tune the band structure in the design of electronic devices [14,19–21]. SW defect is a classical topological defect in graphene and Stone-Wales graphene (SWG) includes two pairs of pentagonal–heptagonal rings. Compared to perfect graphene, SW graphene is more sensitive in absorbing mercaptan, ozone, and formaldehyde [22,23]. The interactions between graphene and different molecules are important for sensor devices based on the graphene. Therefore, experimental or theoretical research has focused on understanding the effect of the electronic and magnetic properties of graphene when molecules adsorb on graphene [24–27]. Transition metals (TMs) absorbed in carbon nanotube and graphene have been given great attention because the absorbed TM atoms can generate novel physical, chemical, and mechanical properties [28–30].

In this study, we used first principle calculations to explore the effect of interactions between SWG and cyclopentadienyl (Cp) or half-metallocene of M (M = Fe, Co, Ni) on their electronic and magnetic properties. Moreover, the geometry and electronic structures were investigated.

2. Calculation Details

Calculations were performed using DFT with van der Waals correctionsas implemented in the CASTEP software of in Material Studio. Generalized gradient approximation (GGA) with the Perdew–Burke–Ernzerhof (PBE) exchange correlation function was used [31]. The SW graphene slab was a 3 × 3 × 1 supercell (9.84 × 9.84 × 15.00 Å, 32 C atoms). The cutoff energy was set to 300 eV. K point of the Brillouin zone and was sampled using 5 × 5 × 1 [32]. The energy convergence standard was 10^{-5} eV per atom during all structural relaxation. The forces on relaxation were less than 0.05 eV/Å. Test calculations which used a higher cutoff energy (400 eV) or a larger K point (7 × 7 × 1) between the SWG sheets were performed and showed less than 4% improvement to the simulation accuracy. Therefore, the calculation parameters were considered to be accurate. We have considered the van der Waals interaction [24,25] between the SWG and Cp. A Cp or half-metallocene was located at the center of the carbon ring of SWG. Considering the structure symmetry of SWG, the hollow sites included the centers of a pentagon ring (H1), hexagon ring (H2), and heptagon ring (H3) (Figure 1).

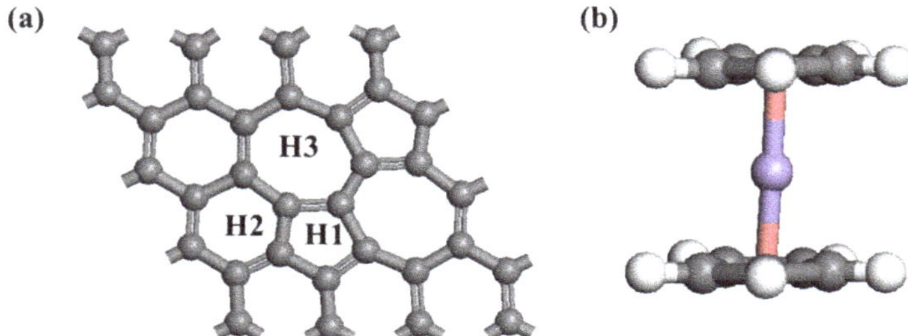

Figure 1. Optimized atomic structures of (**a**) Stone-Wales graphene (SWG) and (**b**) ferrocene.

The adsorption system stability was estimated using the absorption energy E_{ads} defined as follows:

$$E_{ads} = E_T - E_{SWG} - E_{Cp} - \mu_M$$

where E_T is the total energy of the half-metallocene of M (M = Fe, Co, Ni) absorbed SWG; E_{SWG} and E_{Cp} are the energy of pristine SWG and Cp, respectively; and μ_M is the energy of metal ions in the cell.

3. Results and Discussion

3.1. Adsorption Configurations and Energy

3.1.1. Adsorption of Cp in Stone-Wales Graphene (SWG)

To understand how the Cp and half-metallocene are adsorbed in SWG, their adsorption configurations were evaluated and are showed in Figure 1. The adsorption energies and value of charge transfer (Q_e) based on the Mulliken population between SWG and adsorbate corresponding to the different adsorption configurations are listed in Table 1. We discussed the adsorption of Cp on the SWG. The calculated results indicated that Cp absorbed on H1 was the most stable. Meanwhile, the adsorption energy and the smallest distance between Cp and SWG layer is shown in Table 1. The results showed that E_{ads} varied from −1.09 eV to −1.15 eV and H1 had the highest value (−1.15 eV), which was much larger than that adsorbed in perfect graphene. The large E_{ads} indicated that the presence of SW defect affected the adsorption process and enhanced the interaction between Cp and the SWG substrate. Figure 2 shows the electron density difference of SWG and Cp/SWG at H1, which indicated that a strong charge transfer process existed between Cp and SWG. Q_e of SWG at H1 was the largest (0.51 e. In fact, the charge distribution was inhomogeneous in SWG. The pentagon ring had a positive charge, whereas the heptagon ring had a negative charge (Figure 2a). Hence, Cp had one electron (i.e., H1) that was absorbed at the site of the pentagon ring. The result was consistent with the absorption energy.

Table 1. Summary of results for transition metal (TM) atoms adsorbed in SWG. The properties listed are adsorption energy (E_{ads}) and the smallest adatom–carbon distance (d_{AC}).

	d	*E*	Q_e
SWG-Cp			
5	3.901	−1.15	0.51
6	3.489	−1.10	0.31
7	3.413	−1.09	0.36
SWG-Fe-Cp			
5	3.692	−5.54	0.38
6	3.384	−5.63	0.54
7	3.603	−5.72	0.58
SWG-Co-Cp			
5	3.653	−6.12	0.38
6	3.383	−6.21	0.44
7	3.456	−6.67	0.48
SWG-Ni-Cp			
5	3.751	−3.82	0.32
6	3.608	−3.83	0.36
7	3.507	−3.85	0.38

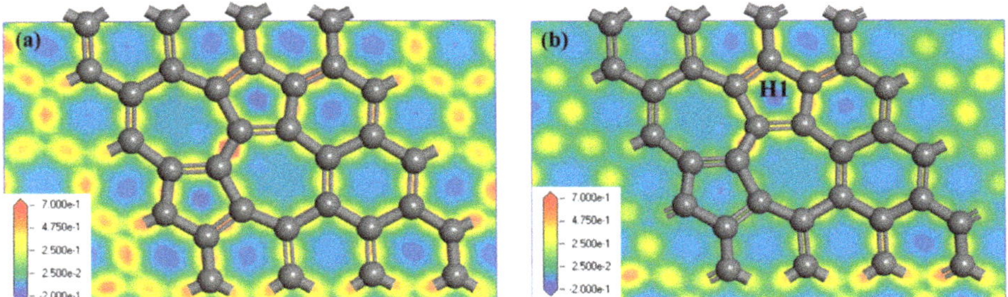

Figure 2. Electron density difference for (**a**) SWG and (**b**) Cyclopentadienyl (Cp)/SWG at the pentagon ring (H1).

3.1.2. Adsorption of Half-Metallocene of *M* (*M* = Fe, Co, Ni) in SWG

In this section, we focus on the adsorption of half-metallocene of *M* (*M* = Fe, Co, Ni) in SWG at three different sites, as shown in Figure 3. The parameters describing the adsorption complexes are depicted in Table 1. The value of E_{ads} ranged from −3.82 eV to −6.23 eV, which was larger than that of SWG. Therefore, metal ions stabilized the adsorption systems of SWG with Cp. The absorption energy had significant differences among different metal ions at the same site. Particularly, SWG-Co-Cp had the strongest E_{ads}, followed by SWG-Fe-Cp and SWG-Ni-Cp. Meanwhile, for the absorption sites, Cp was located closer to the heptagon ring (H3), suggesting that the metal ions played an important role in the adsorption substrate. M^{2+} (*M* = Fe, Co, Ni) with Cp (one electron) had one positive charge, and the heptagon ring had a partial negative charge. Therefore, Cp was located closer to heptagon ring. of SWG at H3 was the largest, which was consistent with its high adsorption energy and value of charge transfer.

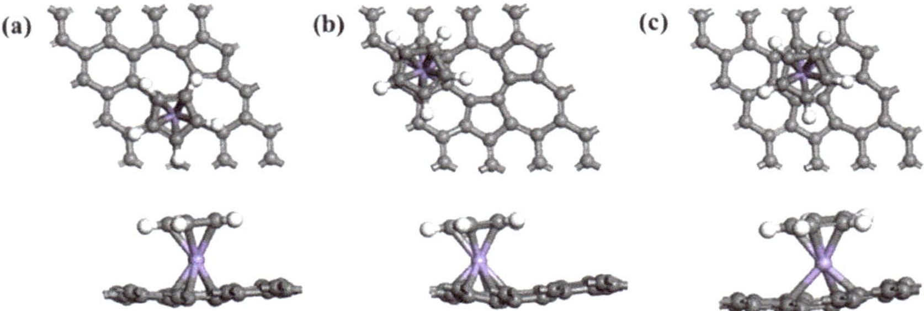

Figure 3. Optimized atomic structures of Cp in SWG at H1 (**a**); hexagon ring (H2) (**b**) and heptagon ring (H3) (**c**).

3.2. Density of States (DOS) of the SWG System

3.2.1. DOS of Cp in SWG

To understand the effect of the electronic properties when Cp absorbed on SWG, the density of states (DOS) of adsorption complexes is illustrated in Figure 4, corresponding to different absorption sites (H1, H2, and H3). The spin-up DOS (majority) and spin-down DOS (minority) were presented in each case, respectively. Pristine SWG is a zero-gap semiconductor in which the Fermi level crossed the Dirac point. In the all cases of absorption configurations, the three systems exhibited metallicity, in which the conduction band passed through the Fermi level. Moreover, the majority and minority DOS were symmetrical, which indicated that their total magnetic moment was zero because of the valence electrons arranged in pairs. From the view point of the molecular orbital, the charge transfer mechanism can be associated with the relative energy positions of the highest occupied molecular orbital (HOMO) and the lowest unoccupied molecular orbital (LUMO) of the adsorbate compared to the SWG Fermi level. If the HOMO is above the Fermi level of SWG, a charge transfer from the adsorbed molecule to SWG may occur. If the LUMO is below the Fermi level, a charge transfer from graphene to the molecule could appear [33]. The HOMO of Cp was 1.68 eV above the Fermi level, deep in the SWG valence band and its LUMO was 7.01 eV above the Fermi level, high in the conduction band of SWG. The HOMO of Cp was the orbital that has a big overlap with the DOS of SWG and thus could cause very big charge transfer value. Therefore, the Cp acted as a very strong donor in the complex systems.

3.2.2. DOS of Half-Metallocene of M (M = Fe, Co, Ni) in SWG

In this part, we focus on the adsorption of half-metallocene of M (M = Fe, Co, Ni) in SWG at three different sites. The DOS and PDOS are shown in Figures 5 and 6. The results indicated the spin-up and spin-down DOS were asymmetrical, which would generate a magnetic moment. Furthermore, the magnetic moment of all complex systems was tuned by the absorption sites and different metal ions. In Cp/Fe/SWG and Cp/Co/SWG, the spin-up DOS value around the Fermi level was more than that of spin-down DOS. In Cp/Ni/SWG, the spin-up DOS value around the Fermi level was less than that of spin-down DOS. In Cp/Ni/SWG, the spin-up DOS around the Fermi level remained zero. Therefore, the zero-gap semiconductor property of the SWG was maintained in the spin-up channel. On the other hand, the spin-down channel of Cp/Ni/SWG showed a non-zero DOS around the Fermi level and metallicity was correspondingly maintained. The interaction between SWG and Fe ion induced effective shifts between the spin-up and spin-down DOS, which lead to a strong magnetic moment.

Considering the strong magnetic property of Fe, Co, and Ni atoms, the spin projected density of states (PDOS) of SWG with half-metallocene of *M* (*M* = Fe, Co, Ni) was showed in Figure 6. The results indicated that the spin DOS of 3d orbital of *M*, Cp, and SWG were all asymmetric between the spin-up and spin-down DOS, which showed that the magnetic property was contributed to by the three parts. The magnetic properties of Cp and SWG were induced from magnetic metal of (*M* = Fe, Co, Ni) by charge transfer. This phenomenon was different from that in a previous study [4].

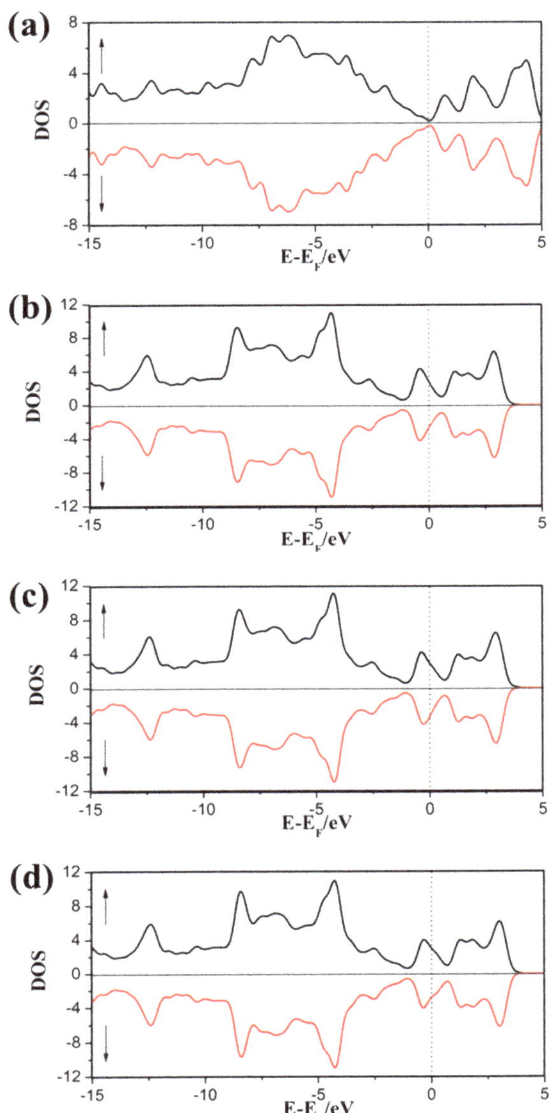

Figure 4. Total electronic density of states (DOS) of pristine SWG (**a**) and Cp on SWG at H1 (**b**); H2 (**c**), and H3 (**d**).

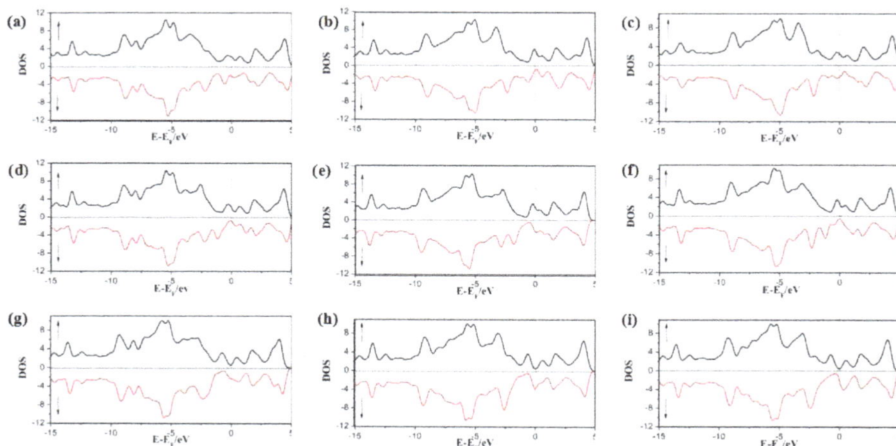

Figure 5. Total electronic DOS of half-metallocene of Fe (**a–c**), Co (**d–e**), and Ni (**g–i**) in SWG at the three hollow sites (H1, H2, and H3).

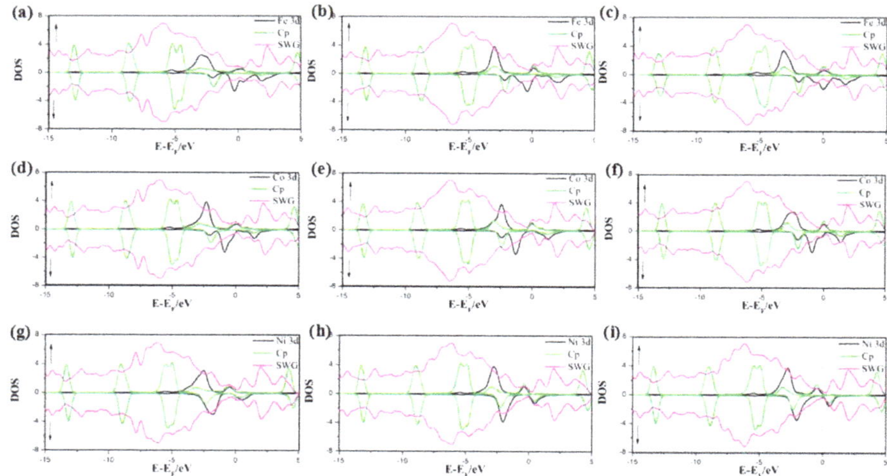

Figure 6. PDOS of half-metallocene of Fe (**a–c**), Co (**d–e**), and Ni (**g–i**) in SWG at the three hollow sites (H1, H2, and H3).

4. Conclusions

The geometrical, energetic, electronic, and magnetic properties of the half-metallocene of *M* (*M* = Fe, Co, Ni) in SWG were investigated using density functional theory (DFT) calculations. The introduction of Cp and half-metallocene increased the conductivity of SWG. Furthermore, the half-metallocene of *M* (*M* = Fe, Co, Ni), with a magnetic behavior, induced different magnetic properties of the adsorption complexes. On the basis of the PDOS results, the magnetic moments of the complex systems were contributed to by the 3d orbital of *M* (*M* = Fe, Co, Ni), molecular orbital of Cp, and SWG. Interestingly, the DOS and magnetic properties were tuned by the absorption sites of Cp and half-metallocene in SWG.

Author Contributions: K.X. and G.Z. conceived and designed the calculated models; K.X., Q.J., X.Z., and L.F. analyzed the data; K.X. wrote the article.

Funding: Support by Lanzhou City University key discipline "Analysis and Treatment of Regional Typical Environmental Pollutants"; the Project of Qinghai Science & Technology Department (No. 2016-ZJ-Y01) and the Open Project of State Key Laboratory of Plateau Ecology and Agriculture, Qinghai University (No. 2017-ZZ-17) is gratefully acknowledged. Computations were done using the National Supercomputing Center in Shenzhen, P. R. China.

Conflicts of Interest: The authors declare no conflicts of interest.

References

1. Khoo, K.H.; Wei, S.T.; John, T.L.Q.; Su, Y. Origin of contact resistance at ferromagnetic metal-graphene interfaces. *ACS Nano* **2016**, *10*, 11219–11227. [CrossRef] [PubMed]
2. Shen, X.; Wang, Z.; Wu, Y.; Liu, X.; He, Y.; Kim, J.K. Multilayer graphene enables higher efficiency in improving thermal conductivities of graphene/epoxy composites. *Nano Lett.* **2016**, *16*, 3585–3593. [CrossRef] [PubMed]
3. Zhou, C.; Szpunar, J.A. Hydrogen storage performance in Pd/graphene nanocomposites. *ACS Appl. Mater. Interface* **2016**, *8*, 25933–25940. [CrossRef] [PubMed]
4. Zhou, Q.; Fu, Z.; Wang, C.; Tang, Y.; Zhang, H.; Yuan, L.; Yang, X. The electronic and magnetic properties of B-doping stone-wales defected graphene decorated with transition-metal atoms. *Physica E* **2015**, *73*, 257–261. [CrossRef]
5. Mohammadi, A.; Haji-Nasiri, S. The electronic transport properties of defected bilayer sliding armchair graphene nanoribbons. *Phys. Lett. A* **2018**, *382*, 1040–1046. [CrossRef]
6. Mombru, D.; Faccio, R.; Mombru, A.W. Possible causes for rippling in a multivacancy graphene system. *Int. J. Quantum Chem.* **2018**, *118*, e25529. [CrossRef]
7. Tran, T.T.; Alotaibi, F.; Nine, M.J.; Silva, R.; Tran, D.N.H.; Janowska, I.; Losic, D. Engineering of highly conductive and ultra-thin nitrogen-doped graphene films by combined methods of microwave irradiation, ultrasonic spraying and thermal annealing. *Chem. Eng. J.* **2018**, *338*, 764–773.
8. Wang, K.; Wang, J.; Wu, Y.; Zhao, S.; Wang, Z.; Wang, S. Nitrogen-doped graphene prepared by a millisecond photo-thermal process and its applications. *Org. Electron.* **2018**, *56*, 221–231. [CrossRef]
9. Yokwana, K.; Ray, S.C.; Khenfouch, M.; Kuyarega, A.T.; Mamba, B.B.; Mhlanga, S.D.; Nxunnalo, E.N. Facile Synthesis of nitrogen doped graphene oxide from graphite flakes and powders: A comparison of their surface chemistry. *J. Nanosci. Nanotechnol.* **2018**, *18*, 5470–5484. [CrossRef] [PubMed]
10. Pellegrino, F.M.D.; Angilella, G.G.N.; Pucci1, R. Effect of impurities in high-symmetry lattice positions on the local density of states and conductivity of grapheme. *Phys. Rev. B* **2009**, *80*, 094203. [CrossRef]
11. Dang, J.S.; Wang, W.W.; Zheng, J.J.; Nagase, S.; Zhao, X. Formation of stone-wales edge: Multistep reconstruction and growth mechanisms of zigzag nanographene. *J. Comp. Chem.* **2017**, *38*, 2241–2247. [CrossRef] [PubMed]
12. Ebrahimi, S.; Azizi, M. The effect of high concentrations and orientations of stone-wales defects on the thermal conductivity of graphene nanoribbons. *Mol. Simulat.* **2018**, *44*, 236–242. [CrossRef]
13. Liu, T.; Zhang, H.; Cheng, X.L.; Xu, Y. Coherent Resonance of quantum plasmons in stone-wales defected graphene-silver nanowire hybrid system. *Front. Phys.* **2017**, *12*, 99–107. [CrossRef]
14. Zaminpayma, E.; Razavi, M.E.; Nayebi, P. Electronic properties of graphene with single vacancy and stone-wales defects. *Appl. Surf. Sci.* **2017**, *414*, 101–106. [CrossRef]
15. Jaskolski, W.; Pelc, M.; Chico, L.; Ayuela, A. Existence of nontrivial topologically protected states at grain boundaries in bilayer graphene: Signatures and electrical switching. *Nanoscale* **2016**, *8*, 6079–6084. [CrossRef] [PubMed]
16. King, R.B. Some Aspects of the symmetry and topology of possible carbon allotrope structures. *J. Math. Chem.* **1998**, *23*, 197–227. [CrossRef]
17. Feng, P.X.L. Tuning in to a graphene oscillator. *Nat. Nanotechnol.* **2013**, *8*, 897–898. [CrossRef] [PubMed]
18. Salary, M.M.; Inampudi, S.; Zhang, K.; Tadmor, E.B.; Mosallaei, H. Mechanical actuation of graphene sheets via optically induced forces. *Phys. Rev. B* **2016**, *94*, 235–403. [CrossRef]
19. Majidi, R.; Tabrizi, K.G. Electronic properties of defect-free and defective bilayer graphene in an electric field. *Nanotubes Carbon Nanostruct.* **2011**, *19*, 532–539. [CrossRef]

20. Wang, Z.Y.; Hu, H.F.; Gu, L.; Wang, W.; Jia, J.F. Electronic and optical properties of zigzag graphene nanoribbon with stone-wales defect. *Acta Phys. Sin.* **2011**, *60*.
21. Zhang, Y.H.; Zhou, K.G.; Xie, K.F.; Gou, X.C.; Zeng, J.; Zhang, H.L.; Peng, Y. Effects of stone-wales defect on the interactions between NH_3, NO_2 and graphene. *J. Nanosci. Nanotechnol.* **2010**, *10*, 7347–7350. [CrossRef] [PubMed]
22. Lee, G.; Lee, B.; Kim, J.; Cho, K. Ozone Adsorption on graphene: Ab initio study and experimental validation. *J. Phys. Chem. C* **2009**, *113*, 14225–14229. [CrossRef]
23. Chen, F.; Tao, N.J. Electron transport in single molecules: From benzene to graphene. *Accounts Chem. Res.* **2009**, *42*, 429–438. [CrossRef] [PubMed]
24. Cox, J.D.; Singh, M.R.; Gumbs, G.; Anton, M.A.; Carreno, F. Dipole-dipole interaction between a quantum dot and a graphene nanodisk. *Phys. Rev. B* **2012**, *86*, 7035. [CrossRef]
25. Kim, S.S.; Kuang, Z.; Ngo, Y.H.; Farmer, B.L.; Naik, R.R. Biotic-abiotic interactions: Factors that influence peptide-graphene interactions. *ACS Appl. Mater. Interface* **2015**, *7*, 20447–20453. [CrossRef] [PubMed]
26. Podeszwa, R. Interactions of graphene sheets deduced from properties of polycyclic aromatic hydrocarbons. *J. Chem. Phys.* **2010**, *132*, 044704. [CrossRef] [PubMed]
27. Zhang, W.; Lu, W.C.; Zhang, H.X.; Ho, K.M.; Wang, C.Z. Hydrogen adatom interaction on graphene: A first principles study. *Carbon* **2018**, *131*, 137–141. [CrossRef]
28. Enriquez, J.I.G.; Villagracia, A.R. Hydrogen adsorption on pristine, defected, and 3d-block transition metal-doped penta-graphene. *Int. J. Hydrog. Energy* **2016**, *41*, 15411. [CrossRef]
29. Gao, J.; Yip, J.; Zhao, J.; Yakobson, B.I.; Ding, F. Graphene nucleation on transition metal surface: Structure transformation and role of the metal step edge. *J. Am. Chem. Soc.* **2012**, *134*, 9534. [CrossRef]
30. Yu, X.; Qu, B.; Zhao, Y.; Li, C.; Chen, Y.; Sun, C.; Gao, P.; Zhu, C. Growth of hollow transition metal (Fe, Co., Ni) oxide nanoparticles on graphene sheets through kirkendall effect as anodes for high-performance lithium-ion batteries. *Chem. Eur. J.* **2016**, *22*, 1638–1645. [CrossRef] [PubMed]
31. Perdew, J.P.; Burke, K.; Ernzerhof, M. Generalized gradient approximation made simple. *Phys. Rev. Lett.* **1996**, *77*, 3865–3868. [CrossRef] [PubMed]
32. Monkhorst, H.J.; Pack, J.D. Special points for brillouin-zone integrations. *Phys. Rev. B* **1977**, *16*, 1748–1749. [CrossRef]
33. Leenaerts, O.; Partoens, B.; Peeters, F.M. Adsorption of H_2O, NH_3, CO, NO_2, and NO on graphene: A first-principles study. *Phys. Rev. B* **2008**, *77*, 125416. [CrossRef]

Article

Optical Properties of Graphene/MoS_2 Heterostructure: First Principles Calculations

Bin Qiu [1], Xiuwen Zhao [1], Guichao Hu [1], Weiwei Yue [1,2], Junfeng Ren [1,2,*] and Xiaobo Yuan [1,*]

1 School of Physics and Electronics, Shandong Normal University, Jinan 250014, China; qiubin@stu.sdnu.edu.cn (B.Q.); zhaoxiuwen@stu.sdnu.edu.cn (X.Z.); hgc@sdnu.edu.cn (G.H.); physics_yue@163.com (W.Y.)

2 Institute of Materials and Clean Energy, Shandong Normal University, Jinan 250014, China

* Correspondence: renjf@sdnu.edu.cn (J.R.); yxb@sdnu.edu.cn (X.Y.)

Received: 29 October 2018; Accepted: 19 November 2018; Published: 21 November 2018

Abstract: The electronic structure and the optical properties of Graphene/MoS_2 heterostructure (GM) are studied based on density functional theory. Compared with single-layer graphene, the bandgap will be opened; however, the bandgap will be reduced significantly when compared with single-layer MoS_2. Redshifts of the absorption coefficient, refractive index, and the reflectance appear in the GM system; however, blueshift is found for the energy loss spectrum. Electronic structure and optical properties of single-layer graphene and MoS_2 are changed after they are combined to form the heterostructure, which broadens the extensive developments of two-dimensional materials.

Keywords: graphene/MoS_2 heterostructure; optical properties; electronic structure

---

1. Introduction

Graphene has been popular among researchers since it was successfully exfoliated by Novoselov et al. in 2004 [1]. Graphene has excellent electrical conductivity [2], excellent mechanical strength [3,4], superior thermal conductivity [5], and high light transmittance in the visible light–infrared area [6]. Graphene has been widely used in applications such as solar cells, lighting, and touch screens [7–14]. However, graphene has been extremely limited in the research and application of some fields because of its zero band gap. One of the methods used to broaden the application of graphene is to form a multilayer structure or heterostructure. Stacking different two-dimensional materials together can form a double-layer or even multi-layer artificial material that is maintained by van der Waals interactions. Such materials are known as van der Waals heterojunctions. Surprising physical properties can be obtained by stacking two-dimensional materials of different properties together. The almost infinitely rich possibilities make the van der Waals heterojunction even more important than the two-dimensional material itself [15–18]. The large surface area, high chemical resistance, high stability, and good electrical conductivity of graphene indicate that graphene sheets are promising as substrates for improving the electrochemical and electrocatalytic properties of metal oxides and metal sulfides. Properties have already been studied in the heterostructure of $Ni(OH)_2$/graphene [19] and SnO_2/graphene [20], which indicates that the heterostructure of graphene also has great research prospects. On the other hand, heterostructures based on graphene and other two-dimensional materials, such as MoS_2, will change their electronic structure and other properties, which has attracted people's attention.

MoS_2 is one of the transition metal dichalcogenides (TMDs). MoS_2 can appear in two-dimensional or three-dimensional forms. The direct band gap will be about 1.8 eV [21,22] when MoS_2 appears as a single-layer two-dimensional material, which makes it a very good semiconductor material. Monolayers of MoS_2 have many excellent properties, such as high electron mobility, low dimensionality, smooth atomic sheet [21,23], and outstanding mechanical properties [24]. Monolayers of MoS_2 have been successfully prepared due to their extraordinary properties [25] and have been extensively studied [21,26–31]. Furthermore, the heterostructure of graphene/MoS_2 opens up possibilities for many applications. For example, Ma et al. [32] systematically investigated the electronic and magnetic properties of perfect, vacancy-doped, and nonmetal elements (H, B, C, N, O, and F) adsorbed $MoSe_2$, $MoTe_2$, and WS_2 monolayers by means of first-principles calculations. In 2011, Chang et al. [33,34] successfully synthesized layered graphene or graphene nanosheet/MoS_2 composites by an L-cysteine-assisted solution-phase methodand the obtained composites showed three-dimensional architecture and excellent electrochemical performances which can act as anode materials for Li-ion batteries. Soon Li et al. [35] developed a selective solvothermal synthesis of MoS_2 nanoparticles on reduced graphene oxide (RGO) sheets and the MoS_2/RGO hybrid exhibited superior electrocatalytic activity in the hydrogen evolution reaction. Coleman et al. [36] showed that hybrid dispersions or composites could be prepared by blending MoS_2 with suspensions of graphene or polymer solutions. A recent study reported the catalytic activity of MoS_2/graphene dots for an oxygen evolution reaction [37]. The above results proved that the heterostructures of GM are useful in applications ranging from electronics to energy storage.

There is still a lack of research of optical properties in GM heterostructures up to now. The heterogeneous structure of graphene has bright prospects of applications and the direct bandgap electronic structure of MoS_2 is an essential property for many optical applications; so, in this paper, we explore the optical properties of GM based on density functional calculations. The structure of this paper is as follows: Section 2 gives the theoretical calculation method, Section 3 gives the result analysis, and Section 4 gives the conclusion.

2. Methods

The DFT calculations we used are performed by the VASP (Vienna ab-initio Simulation Package) software package [38,39]. The lattice constant of the MoS_2 monolayer is 3.16Å, and the lattice constant of pure graphene is 2.47Å, so the supercell of MoS_2 we used was 4*4*1, and the supercell of graphene was 5*5*1. The lattice mismatch ratio of the system was about 2.29%. We stacked monolayer graphene and monolayer MoS_2 to form the heterostructure of GM, which is shown in Figure 1. In order to reduce the interaction between the periodic structures in the vertical direction when constructing the model, a 20Å vacuum is added. In the theoretical calculations, we use the projector-augmented wave (PAW) [40,41] method to describe the interaction between ions and electrons. At the same time, the exchange-correlation potential is selected based on the Generalized Gradient Approximation (GGA [42]) in terms of the Perdew–Burke–Ernzerhof (PBE [42]) functional, which is often used to calculate the molecular adsorption at the electrode surface. The cutting power of the plane wave is set to 500 eV. When the structure relaxes, the convergence precision of each interatomic force is 0.02 eV/Å, and the self-consistent convergence energy is not higher than 10^{-4} eV. The Brillouin zone was summed according to the $9\times9\times1$ Monkhorst–Pack characteristic K point. Based on the above conditions, the calculated distance between graphene and MoS_2 is 3.64Å. Then, the electronic structure and the optical properties of the heterostructures are calculated. Van der Waals interactions are included in the calculations.

The optical properties can be modeled by the dielectric constant of the system. We use the superposition of Lorentz oscillators to model the complex dielectric function $\varepsilon(\omega) = \varepsilon_1(\omega) + i\varepsilon_2(\omega)$ of the heterostructure, which is a function of photon energy. Generally speaking, the dielectric constant is the real part of the complex permittivity, $\varepsilon_1(\omega)$. The dielectric constant is caused by various kinds of displacement polarization inside the material and represents the energy storage term of the material. The imaginary part of the complex permittivity, $\varepsilon_2(\omega)$, is related to the absorption (loss or gain) of the material. The steering polarization can not keep up with the various relaxation polarizations caused by the change of the external high-frequency electric field, and represents the loss term of the material. The formula of $\varepsilon_2(\omega)$ is as follows:

$$\varepsilon_2(\omega) = \frac{4\pi^2 e^2}{\Omega} \lim_{q\to 0} \frac{1}{q^2} \sum_{c,v,k} 2w_k \delta(\in_{ck} - \in_{vk} - w) \times \langle u_{ck} + e_{\alpha q} | u_{vk} \rangle \langle u_{ck} + e_{\beta q} | u_{vk} \rangle^* \quad (1)$$

The real part $\varepsilon_1(\omega)$ of the dielectric function can be obtained by using the Kramers–Kroing relation,

$$\varepsilon_1(\omega) = 1 + \frac{2}{\pi} P \int_0^\infty \frac{\varepsilon_2^{\alpha\beta}(w')w'}{w'^2 - w^2 + i\eta} d\omega' \quad (2)$$

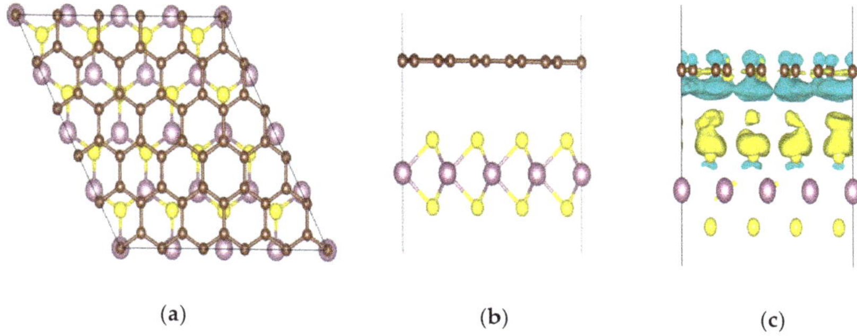

(a) (b) (c)

Figure 1. Top (**a**) and side (**b**) views of the Graphene/MoS_2 (GM) heterostructure. (**c**) The differential charge density distributions of GM. Gray, purple, and yellow atoms represent C, Mo, and S atoms, respectively. Blue means loss electrons and yellow means gain electrons.

Other optical constants can also be obtained from the dielectric function. For example, the absorption coefficient $\alpha(\omega)$, refractive index $n(\omega)$, reflectance $R(\omega)$, and energy loss spectrum $L(\omega)$ can all be derived by $\varepsilon_1(\omega)$ and $\varepsilon_2(\omega)$. The formulas are:

$$\alpha(\omega) = \frac{\sqrt{2}\omega}{c} \left\{ \left[\varepsilon_1^2(\omega) + \varepsilon_2^2(\omega) \right]^{\frac{1}{2}} - \varepsilon_1(\omega) \right\}^{\frac{1}{2}} \quad (3)$$

$$n(\omega) = \frac{1}{\sqrt{2}} \left\{ \left[\varepsilon_1^2(\omega) + \varepsilon_2^2(\omega) \right]^{\frac{1}{2}} + \varepsilon_1(\omega) \right\}^{\frac{1}{2}} \quad (4)$$

$$R(\omega) = \left| \frac{\sqrt{\varepsilon_1(\omega) + i\varepsilon_2(\omega)} - 1}{\sqrt{\varepsilon_1(\omega) + i\varepsilon_2(\omega)} + 1} \right|^2 \quad (5)$$

$$L(\omega) = \frac{\varepsilon_2(\omega)}{\varepsilon_1^2(\omega) + \varepsilon_2^2(\omega)} \quad (6)$$

3. Results and Discussion

In order to illustrate the similarities and the differences of the graphene monolayer, MoS_2 monolayer, and the GM heterostructure, we first calculate the electronic structures of the three systems. The energy band structures and the electronic density of states (DOS) for the three systems are shown in Figure 2. It can be found that our calculated curves are well matched with the results of previous calculations [43,44]. As shown in Figure 2, graphene is a zero bandgap material and MoS_2 is a material with a band gap of 1.73 eV. After they are stacked together to form the GM structure, as shown in Figure 1, the band gap is 3.49 meV for GM heterostructures, which can be obtained from the embedded figure in Figure 2c. Based on the interlayer interactions between G and

M, there will be a change in the on-site energy of atoms in the G layer, so the band gap opens [44,45]. The upward shift of the Dirac point of graphene with respect to the Fermi level indicates that holes are donated by the MoS_2 monolayer, which can be confirmed by the charge transfer between graphene and MoS_2 after stacking. Figure 1c gives the differential charge density distributions, blue means loss electrons and yellow means gain electrons. It is clear from the figure that holes in G are donated by M monolayer after the stacking. From Figure 2 we can clearly see that after the heterostructure is formed, the electronic structure changes greatly. Therefore, we speculate that the formation of the GM heterostructure will influence the optical properties compared with single-layer graphene or MoS_2.

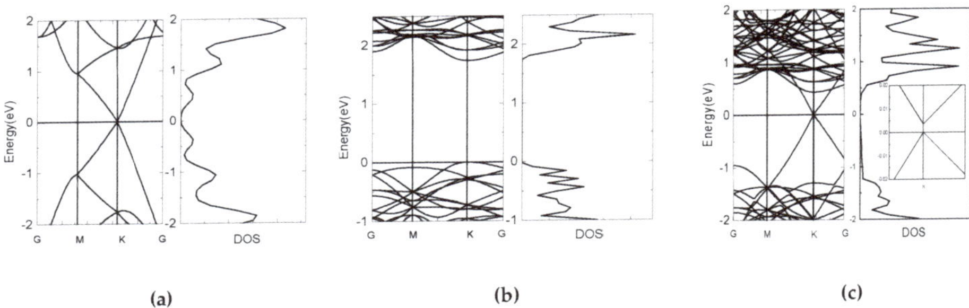

Figure 2. Band structure and density of states (DOS) of graphene (**a**), MoS_2 (**b**) and the GM heterostructure (**c**), respectively. The embedded figure in (**c**) shows the zoom in of the band structure near Fermi Energy.

The calculated dielectric constants $\varepsilon(\omega)$ of the monolayer graphene (G), monolayer MoS_2 (M), and GM heterostructure are shown in Figure 3. Figure 3a shows the parallel direction of the $\varepsilon_1(\omega)$. We can clearly see from the figure that the overall trends for all systems are almost identical with only small differences. In fact, people are more interested in the changes that occur in the visible light region. In the visible light region, the value of the $\varepsilon_1(\omega)$ is obviously the largest in the GM system, followed by the M system, and finally the G system. Comparing the GM and G system at the low-energy zone, it can be found that the parallel direction of $\varepsilon_1(\omega)$ for the two systems not only changes at the maximum values, but also GM has an obvious blueshift of $\varepsilon_1(\omega)$ relative to the G system. Figure 3b shows the $\varepsilon_1(\omega)$ in the vertical direction and we find similar regularity with those of Figure 3a. Under the same analysis of the three systems in the low-energy region, we find that the most obvious change is a more obvious redshift for the GM system compared with the G system. This is because the GM system is an anisotropic material and the parallel direction of the $\varepsilon_1(\omega)$ illustrates differences in the vertical and horizontal directions. Figure 3c, d shows the parallel and vertical directions of the imaginary part of the dielectric constant, respectively. Same properties between the real and the imaginary parts of $\varepsilon_1(\omega)$ can be found. The peak value of the dielectric constant of the GM system has been significantly improved compared with G and M and different degrees of redshift or blueshift can also be found.

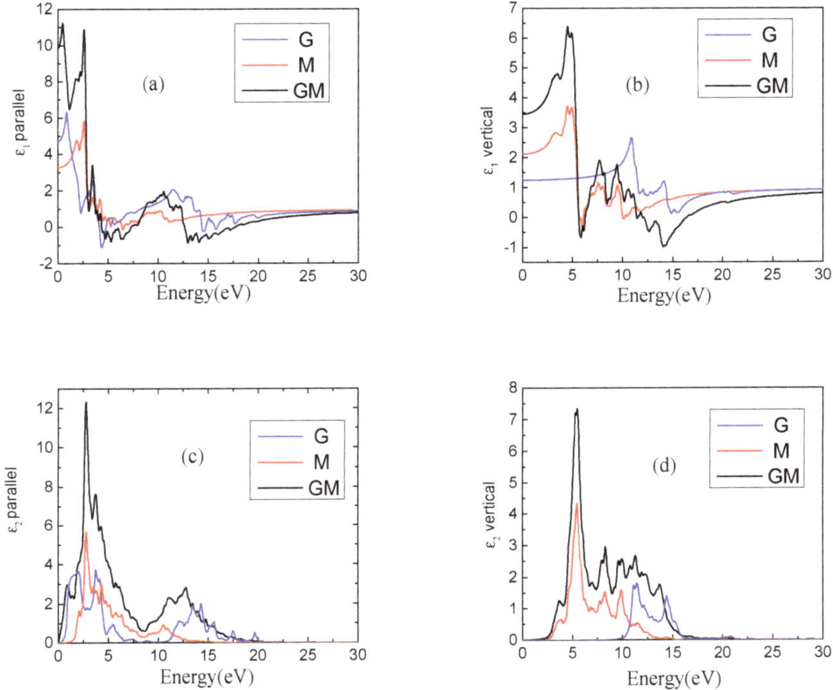

Figure 3. The complex dielectric constants of monolayer graphene (G), monolayer MoS_2 (M) and GM systems. (**a**,**b**) represent the parallel and vertical components of the real part of the dielectric constant, (**c**,**d**) represent parallel and vertical components of the imaginary part of the dielectric constant, respectively.

Figure 4a shows the absorption coefficient $\alpha(\omega)$ in the parallel direction. The overall change trend of the GM and M systems are similar, and the only difference is in the peak values. There are obvious differences between the GM and G systems. The $\alpha(\omega)$ of the GM system is more volatile than the G system at the peak position. Among the three systems, GM usually has a large $\alpha(\omega)$ value in most cases; however, in the visible region, G is slightly larger than that of the GM system. A zoom in the region between 0 and 2 eV of Figure 4a is also embedded. The value of the intersection between the reverse tangent and the x-axis is the optical band gap in the region between 0 and 2 eV in Figure 4a. It can be found from Figure 4a, that the optical band gap of G is about 0.75 eV, and the optical band gap of M is about 1.63 eV. However, the band gaps are around 0.41 eV and 1.40 eV when the system is going from G and M to GM. It is well known that a photoelectron can be excited with less energy when the optical band gap is small. The optical band gap of GM is significantly reduced, which indicates that we can use a lower energy to excite a photoelectron in GM compared with the G and M systems. The vertical direction of $\alpha(\omega)$ is given in Figure 4b. The overall change trend of the vertical direction, $\alpha(\omega)$, has a similar regularity compared with the parallel direction. The obvious difference is that the GM system has a large redshift in the vertical direction compared with the G system. The $\alpha(\omega)$ is greatly improved for the GM system compared with the G and M systems, so the GM system is indeed superior to the G and M systems in terms of absorption properties.

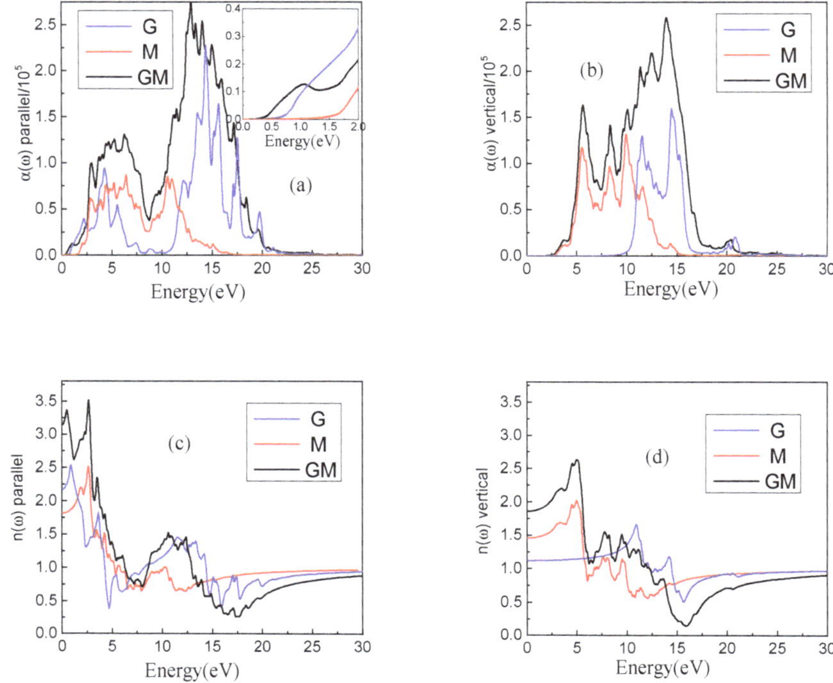

Figure 4. The absorption coefficient α(ω) and the refractive index n(ω) of three systems. (**a**,**b**) represent parallel and vertical components of the absorption coefficient α(ω), (**c**,**d**) represent parallel and vertical components of the refractive index n(ω), respectively.

The parallel direction and vertical direction of the refractive index n(ω) are given in Figure 4c,d, respectively. According to the formula for calculating the refractive index, i.e., Equation (4), we can see that the refractive index is essentially related to the real and the imaginary parts of the dielectric constant. By comparing the dielectric constant of Figure 3 and the refractive index image of Figure 4, it can be found that the change trends of Figure 3a,b are similar with those in Figure 4c,d, which means that the effects of the real part of the dielectric constant on the refractive index play the leading role. We found that the n(ω), especially in the visible light range, has a large value for the GM system. The heat preservation characteristics will be good if the material has a big refractive index. This property can be applied to materials that require constant temperature conditions.

The parallel and vertical directions of the reflectance R(ω) are given in Figure 5a,b, respectively. For parallel directions, the GM system is significantly higher than those of the G and M systems, especially in the visible region. It is obvious that the GM system has a certain redshift relative to the G system, and this phenomenon is also reflected in the vertical direction. In the visible light region, the value of the GM system is also higher than those of the other two systems.

The energy loss spectra L(ω) are given in Figure 5c,d. In the parallel direction, L(ω) of the GM in the low-energy region is significantly less than those of the other two systems. Especially for the G system, the maximum energy loss in the low-energy zone reaches 2, while the GM system is around 0.3. As the energy increases, energy losses also increase. The energy loss of the GM system is concentrated inthe range of 15–20 eV, however for the G and M systems, the energy losses are concentrated in the range of 5–20 eV and they span a large energy extent. In the vertical direction, the energy losses of the three systems in the low-energy region are almost zero, indicating that the loss of power in the vertical direction is small in the low-energy region. The energy loss of the GM system is almost concentrated between 15 eV and 18 eV, while the energy of the G and the M system are lost relatively evenly between 5 eV and 15 eV, which means that the ability to control the energy loss of the GM system is the best. In addition, the GM system is relatively blueshifted for both horizontal and vertical energy loss compared with the G and M systems.

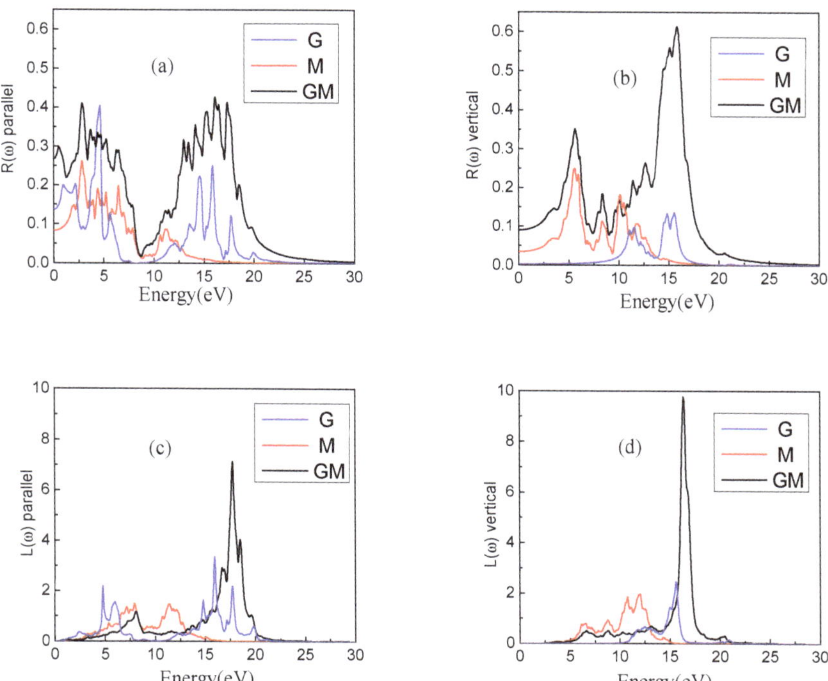

Figure 5. The reflectance R(ω) and the energy loss spectrum L(ω) of three systems. (**a**,**b**) represent parallel and vertical components of the reflectance R(ω), (**c**,**d**) represent parallel and vertical components of the energy loss spectrum L(ω), respectively.

4. Conclusions

In this article, we mainly discuss the electronic structure and the optical properties of GM heterostructures from the first principles calculations. Based on the DFT theory, dielectric constant, ε(ω) absorption coefficient α(ω), refractive index n(ω), reflectivity R(ω), and energy loss spectrum L(ω) of the systems are calculated. It is found that there is indeed a clear improvement of the optical properties for the GM system comparedto the G and M systems. The band gap and the dielectric constants become large for the GM system and there are redshifts for the absorption coefficient, refractive index, and the reflectance. A blueshift is found for the energy loss spectrum in the GM system. All of the above results show that, due to the formation of the heterojunctions, the optical properties of the GM system have been significantly improved compared with the single layers, which deliversa more effective way to use two-dimensional materials in optical applications.

Author Contributions: B.Q. did the calculations and wrote the paper, X.Z. collected the references, G.H. prepared the figures, W.Y. and J.R. analyzed the data, X.Y. generated the research idea. All authors read and approved the final manuscript.

Funding: This research was funded by the National Natural Science Foundation of China (Grant No. 11674197) and the Natural Science Foundation of Shandong Province (Grant Nos.ZR2018MA042).

Acknowledgments: This work was supported by the Taishan Scholar Project of Shandong Province.

Conflicts of Interest: The authors declare no conflicts of interest.

References

1. Novoselov, K.S.; Geim, A.K.; Morozov, S.V.; Jiang, D.; Zhang, Y.; Dubonos, S.V.; Grigorieva, I.V.; Firsov, A.A. Electric field effect in atomically thin carbon films. *Science* **2004**, *306*, 666–669. [CrossRef] [PubMed]
2. Novoselov, K.S.; Geim, A.K.; Morozov, S.V.; Jiang, D.; Katsnelson, M.I.; Grigorieva, I.V.; Dubonos, S.V.; Firsov, A.A. Two-dimensional gas of massless Dirac fermions in graphene. *Nature* **2005**, *438*, 197. [CrossRef] [PubMed]
3. Park, H.; Chang, S.; Zhou, X.; Kong, J.; Palacios, T.; Gradecak, S. Flexible graphene electrode-based organic photovoltaics with record-high efficiency. *Nano Lett.* **2014**, *14*, 5148–5154. [CrossRef] [PubMed]
4. Lee, C.; Wei, X.; Kysar, J.W.; Hone, J. Measurement of the elastic properties and intrinsic strength of monolayer graphene. *Science* **2008**, *321*, 385–388. [CrossRef] [PubMed]
5. Hu, Y.H.; Wang, H.; Hu, B. Thinnest two-dimensional nanomaterial-graphene for solar energy. *ChemSusChem* **2010**, *3*, 782–796. [CrossRef] [PubMed]
6. Nair, R.R.; Blake, P.; Grigorenko, A.N.; Novoselov, K.S.; Booth, T.J.; Stauber, T.; Peres, N.M.R.; Geim, A.K. Fine Structure Constant Defines Visual Transparency of Graphene. *Science* **2008**, *320*. [CrossRef] [PubMed]
7. Wang, Y.; Qu, Z.; Liu, J.; Tsang, Y.H. Graphene Oxide Absorbers for WattLevel High-Power Passive Mode-Locked Nd:GdVO Laser Operating at 1 μm. *J. Lightwave Technol.* **2012**, *30*, 3259–3262. [CrossRef]
8. Zhu, H.; Liu, J.; Jiang, S.; Xu, S.; Su, L.; Jiang, D.; Qian, X.; Xu, J. Diode-pumped Yb,Y:CaF2 laser mode-locked by monolayer graphene. *Opt. Laser Technol.* **2015**, *75*, 83–86. [CrossRef]
9. Kurapati, R.; Kostarelos, K.; Prato, M.; Bianco, A. Biomedical Uses for 2D Materials Beyond Graphene: Current Advances and Challenges Ahead. *Adv. Mater.* **2016**, *28*, 6052–6074. [CrossRef] [PubMed]
10. Chen, X.; Meng, R.; Jiang, J.; Liang, Q.; Yang, Q.; Tan, C.; Sun, X.; Zhang, S.; Ren, T. Electronic structure and optical properties of graphene/staneneheterobilayer. *Phys. Chem. Chem. Phys.* **2016**, *18*, 16302–16309. [CrossRef] [PubMed]
11. Rezania, H.; Yarmohammadi, M. The effects of impurity doping on the optical properties of biased bilayer graphene. *Opt. Mater.* **2016**, *57*, 8–13. [CrossRef]
12. Nelson, F.J.; Kamineni, V.K.; Zhang, T.; Comfort, E.S.; Lee, J.U.; Diebold, A.C. Optical properties of large-area polycrystalline chemical vapor deposited graphene by spectroscopic ellipsometry. *Appl. Phys. Lett.* **2010**, *97*, 3. [CrossRef]
13. Ren, Y.; Brown, G.; Mary, R.; Demetriou, G.; Popa, D.; Torrisi, F.; Ferrari, A.C.; Chen, F.; Kar, A.K. 7.8-GHz Graphene-Based 2-μm Monolithic Waveguide Laser. *IEEE J. Sel. Top. QuantumElectron.* **2015**, *21*, 395–400. [CrossRef]
14. Weber, J.W.; Calado, V.E.; van de Sanden, M.C.M. Optical constants of graphene measured by spectroscopic ellipsometry. *Appl. Phys. Lett.* **2010**, *97*, 091904. [CrossRef]
15. Frisenda, R.; Navarro-Moratalla, E.; Gant, P.; Perez De Lara, D.; Jarillo-Herrero, P.; Gorbachev, R.V.; Castellanos-Gomez, A. Recent progress in the assembly of nanodevices and van der Waals heterostructures by deterministic placement of 2D materials. *Chem. Soc. Rev.* **2018**, *47*, 53–68. [CrossRef] [PubMed]
16. Frisenda, R.; Molina-Mendoza, A.J.; Mueller, T.; Castellanos-Gomez, A.; van der Zant, H.S.J. Atomically thin p-n junctions based on two-dimensional materials. *Chem. Soc. Rev.* **2018**, *47*, 3339–3358. [CrossRef] [PubMed]
17. Novoselov, K.S.; Mishchenko, A.; Carvalho, A.; Castro Neto, A.H. 2D materials and van der Waals heterostructures. *Science* **2016**, *353*. [CrossRef] [PubMed]
18. Liu, Y.; Weiss, N.O.; Duan, X.; Cheng, H.-C.; Huang, Y.; Duan, X. Van der Waals heterostructures and devices. *Nat. Rev. Mater.* **2016**, *1*. [CrossRef]
19. Wang, H.; Casalongue, H.S.; Liang, Y.; Dai, H. $Ni(OH)_2$ Nanoplates Grown on Graphene as Advanced Electrochemical Pseudocapacitor Materials. *J. Am. Chem. Soc.* **2010**, *132*, 7472–7477. [CrossRef] [PubMed]
20. Wang, X.; Zhou, X.; Yao, K.; Zhang, J.; Liu, Z. A SnO_2/graphene composite as a high stability electrode for lithium ion batteries. *Carbon* **2011**, *49*, 133–139. [CrossRef]
21. Mak, K.F.; Lee, C.; Hone, J.; Shan, J.; Heinz, T.F. Atomically thin MoS(2): A new direct-gap semiconductor. *Phys. Rev. Lett.* **2010**, *105*, 136805. [CrossRef] [PubMed]

22. Splendiani, A.; Sun, L.; Zhang, Y.; Li, T.; Kim, J.; Chim, C.Y.; Galli, G.; Wang, F. Emerging photoluminescence in monolayer MoS_2. *Nano Lett.* **2010**, *10*, 1271–1275. [CrossRef] [PubMed]
23. Lin, M.-W.; Liu, L.; Lan, Q.; Tan, X.; Dhindsa, K.S.; Zeng, P.; Naik, V.M.; Cheng, M.M.-C.; Zhou, Z. Mobility enhancement and highly efficient gating of monolayer MoS_2transistors with polymer electrolyte. *J. Phys. D Appl. Phys.* **2012**, *45*, 345102. [CrossRef]
24. Lorenz, T.; Ghorbani-Asl, M.; Joswig, J.O.; Heine, T.; Seifert, G. Is MoS(2) a robust material for 2D electronics? *Nanotechnology* **2014**, *25*, 445201. [CrossRef] [PubMed]
25. Kim, D.; Sun, D.; Lu, W.; Cheng, Z.; Zhu, Y.; Le, D.; Rahman, T.S.; Bartels, L. Toward the growth of an aligned single-layer MoS_2 film. *Langmuir* **2011**, *27*, 11650–11653. [CrossRef] [PubMed]
26. Han, S.W.; Kwon, H.; Kim, S.K.; Ryu, S.; Yun, W.S.; Kim, D.H.; Hwang, J.H.; Kang, J.S.; Baik, J.; Shin, H.J.; et al. Band-gap transition induced by interlayer van der Waals interaction in MoS_2. *Phys. Rev. B* **2011**, *84*. [CrossRef]
27. Ataca, C.; Topsakal, M.; Akturk, E.; Ciraci, S. A Comparative Study of Lattice Dynamics of Three- and Two-Dimensional MoS_2. *J. Phys. Chem. C* **2011**, *115*, 16354–16361. [CrossRef]
28. Lebègue, S.; Eriksson, O. Electronic structure of two-dimensional crystals fromabinitiotheory. *Phys. Rev. B* **2009**, *79*. [CrossRef]
29. Yu, S.; Xiong, H.D.; Eshun, K.; Yuan, H.; Li, Q. Phase transition, effective mass and carrier mobility of MoS_2 monolayer under tensile strain. *Appl. Surf. Sci.* **2015**, *325*, 27–32. [CrossRef]
30. Ma, D.; Ju, W.; Li, T.; Zhang, X.; He, C.; Ma, B.; Tang, Y.; Lu, Z.; Yang, Z. Modulating electronic, magnetic and chemical properties of MoS_2 monolayer sheets by substitutional doping with transition metals. *Appl. Surf. Sci.* **2016**, *364*, 181–189. [CrossRef]
31. Ma, D.; Ju, W.; Li, T.; Zhang, X.; He, C.; Ma, B.; Lu, Z.; Yang, Z. The adsorption of CO and NO on the MoS_2 monolayer doped with Au, Pt, Pd, or Ni: A first-principles study. *Appl. Surf. Sci.* **2016**, *383*, 98–105. [CrossRef]
32. Ma, Y.; Dai, Y.; Guo, M.; Niu, C.; Lu, J.; Huang, B. Electronic and magnetic properties of perfect, vacancy-doped, and nonmetal adsorbed $MoSe_2$, $MoTe_2$ and WS_2 monolayers. *Phys. Chem. Chem. Phys.* **2011**, *13*, 15546–15553. [CrossRef] [PubMed]
33. Chang, K.; Chen, W. In situ synthesis of MoS_2/graphene nanosheet composites with extraordinarily high electrochemical performance for lithium ion batteries. *Chem. Commun.* **2011**, *47*, 4252–4254. [CrossRef] [PubMed]
34. Chang, K.; Chen, W. L-Cysteine-Assisted Synthesis of Layered MoS_2/Graphene Composites with Excellent Electrochemical Performances for Lithium Ion Batteries. *ACS Nano* **2011**, *5*, 4720. [CrossRef] [PubMed]
35. Li, Y.; Wang, H.; Xie, L.; Liang, Y.; Hong, G.; Dai, H. MoS_2 nanoparticles grown on graphene: An advanced catalyst for the hydrogen evolution reaction. *J. Am. Chem. Soc.* **2011**, *133*, 7296–7299. [CrossRef] [PubMed]
36. Coleman, J.N. Two-Dimensional Nanosheets Produced by Liquid Exfoliation of Layered Materials. *Science* **2011**, *331*, 568–571. [CrossRef] [PubMed]
37. Mohanty, B.; Ghorbani-Asl, M.; Kretschmer, S.; Ghosh, A.; Guha, P.; Panda, S.K.; Jena, B.; Krasheninnikov, A.V.; Jena, B.K. MoS_2 Quantum Dots as Efficient Catalyst Materials for the Oxygen Evolution Reaction. *ACS Catal.* **2018**, *8*, 1683–1689. [CrossRef]
38. Kresse, G.; Furthmüller, J. Efficiency of Ab-Initio Total Energy Calculations for Metals and Semiconductors Using a Plane-Wave Basis Set. *Comput. Mater. Sci.* **1996**, *6*, 15–50. [CrossRef]
39. Kresse, G.; Furthmüller, J. Efficient iterative schemes for ab initio total-energy calculations using a plane-wave basis set. *Phys. Rev. B* **1996**, *54*, 11169–11186. [CrossRef]
40. Kresse, G.; Joubert, D. From ultrasoft pseudopotentials to the projector augmented-wave method. *Phys. Rev. B* **1999**, *59*, 1758–1775. [CrossRef]
41. Blöchl, P.E. Projector augmented-wave method. *Phys. Rev. B* **1994**, *50*, 17953–17979. [CrossRef]
42. Perdew, J.P.; Burke, K.; Ernzerhof, M. Generalized Gradient Approximation Made Simple. *Phys. Rev. Lett.* **1996**, *77*, 3865–3868. [CrossRef] [PubMed]
43. Phuc, H.V.; Hieu, N.N.; Hoi, B.D.; Phuong, L.T.; Nguyen, C.V. First principle study on the electronic properties and Schottky contact of graphene adsorbed on MoS_2 monolayer under applied out-plane strain. *Surf. Sci.* **2018**, *668*, 23–28. [CrossRef]

44. Ma, Y.; Dai, Y.; Guo, M.; Niu, C.; Huang, B. Graphene adhesion on MoS(2) monolayer: An ab initio study. *Nanoscale* **2011**, *3*, 3883–3887. [CrossRef] [PubMed]
45. Ghorbani-Asl, M.; Bristowe, P.D.; Koziol, K.; Heine, T.; Kuc, A. Effect of compression on the electronic, optical and transport properties of MoS_2/graphene-based junctions. *2D Mater.* **2016**, *3*, 025018. [CrossRef]

Article

Graphene Schottky Junction on Pillar Patterned Silicon Substrate

Giuseppe Luongo [1,2,*], Alessandro Grillo [1], Filippo Giubileo [2], Laura Iemmo [1,2], Mindaugas Lukosius [3], Carlos Alvarado Chavarin [3], Christian Wenger [3,4] and Antonio Di Bartolomeo [1,2,5,*]

1 Physics Department "E. R. Caianiello", University of Salerno, via Giovanni Paolo II n. 132, 84084 Fisciano, Italy; agrillo@unisa.it (A.G.); liemmo@unisa.it (L.I.)
2 CNR-SPIN Salerno, via Giovanni Paolo II n. 132, 84084 Fisciano, Italy; filippo.giubileo@spin.cnr.it
3 IHP–Leibniz Institut fuer innovative Mikroelektronik, Im Technologiepark 25, 15236 Frankfurt (Oder), Germany; lukosius@ihp-microelectronics.com (M.L.); alvarado@ihp-microelectronics.com (C.A.C.); wenger@ihp-microelectronics.com (C.W.)
4 Brandenburg Medical School Theodor Fontane, 16816 Neuruppin, Germany
5 Interdepartmental Centre NanoMates, University of Salerno, via Giovanni Paolo II n. 132, 84084 Fisciano, Italy
* Correspondence: giluongo@unisa.it (G.L.); adibartolomeo@unisa.it (A.D.B.); Tel.: +39-089-96-8131 (G.L.); +39-089-96-9189 (A.D.B.)

Received: 29 March 2019; Accepted: 22 April 2019; Published: 26 April 2019

Abstract: A graphene/silicon junction with rectifying behaviour and remarkable photo-response was fabricated by transferring a graphene monolayer on a pillar-patterned Si substrate. The device forms a 0.11 eV Schottky barrier with 2.6 ideality factor at room temperature and exhibits strongly bias- and temperature-dependent reverse current. Below room temperature, the reverse current grows exponentially with the applied voltage because the pillar-enhanced electric field lowers the Schottky barrier. Conversely, at higher temperatures, the charge carrier thermal generation is dominant and the reverse current becomes weakly bias-dependent. A quasi-saturated reverse current is similarly observed at room temperature when the charge carriers are photogenerated under light exposure. The device shows photovoltaic effect with 0.7% power conversion efficiency and achieves 88 A/W photoresponsivity when used as photodetector.

Keywords: graphene; Schottky barrier; diode; photodetector; heterojunction; MOS (Metal Oxide Semiconductor) capacitor; responsivity

---

1. Introduction

The discovery of two-dimensional (2D) materials such as graphene [1], MoS_2 [2,3], WSe_2 [4,5], phosphorene and so on [6], has attracted the interests of the scientific community in the recent years. Graphene is still one the most studied materials for its 2D honeycomb structure, high electron mobility, high electrical and thermal conduction, low optical absorption coefficient and easy fabrication methods [1,7,8]. Large graphene layers can be easily synthesized by chemical vapor deposition (CVD) and integrated into the existing semiconductor device technologies. These properties make graphene the perfect candidate to realize a new generation of transistors [9–14], diodes [15–20], chemical-biological sensors [21–23], photodetectors and solar cells [24–30]. In the recent years, a lot of activity has been focused on the graphene/silicon junction (gr/Si) as one of the simplest graphene devices offering the possibility to study the physical phenomena that occur at the interface between 2D and 3D materials [31]. The gr/Si junction usually forms a Schottky barrier and behaves as a rectifier with a current-voltage (I-V) characteristic similar to that of a metal/semiconductor Schottky diode [31,32]. Because of its particular band structure, graphene possesses low electron density of states close to the Dirac point, hence the Fermi level is highly dependent on charge transfer to or from it. In the gr/Si junction, the application of a bias affects the charge transfer process and the consequent shift of the graphene Fermi energy modulates the gr/Si Schottky barrier height, which becomes therefore bias dependent [31,32]. Indeed, adding such a feature into the standard thermionic emission (T.E.) theory provides an accurate model to describe the gr/Si experimental I-V characteristics [31,33]. Gr/Si Schottky diodes are characterized by a higher ideality factor ($n > 2$) than metal/semiconductor devices ($n \sim 1.3$) [31]. The higher n arises because native oxide layers are generally formed at the interface during the graphene transfer process along with silicon interface trap states and/or metallic contamination [34,35]. Obviously, the ideality of the junction can be improved by reducing the interface defects, for instance through a suitable patterning of the substrate. Indeed, the gr/Si tip junctions that we presented in a previous work showed an ideality factor of 1.5 as the patterning of the Si substrate in a tip-array geometry reduces the probability of finding defects or contaminates at the junction, compared to a planar junction of the same area [17]. In addition to that, the tip geometry amplifies the electric field close to the junction, inducing a potential that shifts the graphene Fermi level even at low bias. We exploited such a feature to realize a bias-tunable graphene-based Schottky barrier device [17].

Modifying the substrate geometry is a viable approach to improve the gr/Si device performance or its photoresponse when used as a photodetector. We remark that the photoresponsivity of the gr/Si junction has been also improved by acting on the device structure. One possible way is to reduce the oxide layer underneath the graphene in order to create a metal/oxide/semiconductor (MOS) capacitor next to the gr/Si junction perimeter. Indeed, such an MOS capacitor plays an important role in the photo-charge collection process, by providing photogenerated carriers from the Si substrate to the junction [16,18–20,36].

In this work, we combine the tip geometry and the MOS capacitor approach, by fabricating a graphene/silicon junction on Si pillars to realize a bias-tunable Schottky diode that can be used also for photovoltaic and photodetection applications. The pillar perimeter works similarly to the nanotips in enhancing the electric field at the junction but is easier to fabricate and provides a better control of the MOS capacitor areas. We present an extensive analysis of the I-V characteristics of gr/Si pillar junction and evaluate the relevant parameters using the T.E. model and the Cheung and Cheung (C.C.) method [31,32,37]. We also investigate the photo response and the photovoltaic effect of the device using white LED light at different intensities.

2. Materials and Methods

Figure 1a shows the schematic view of the gr/Si-pillar junction. Starting from a highly n-doped silicon substrate ($\sim 10^{18}$ cm^{-3}) three pillars with the height of ~500 nm and square sections of 30 μm, 50 μm and 100 μm per side were patterned by photolithography (Figure 1b). In a gr/Si junction the Schottky barrier is controlled by the sharper geometries, that is by the pillar perimeter in our case. As the three pillars have similar perimeter/area ratio (~10%), we expect that they contribute in a similar way to the junction properties. A SiO_2 layer was CVD-deposited until it covered the silicon pillars. Chemical-mechanical polishing (CMP) was then used to remove the oxide layer on the top of the pillars. After that, a graphene layer was transferred from Cu foil on the pillars with a method detailed elsewhere [35].

The Raman spectrum of the graphene measured on the SiO_2 and Si pillars is shown in Figure 1c. The plot shows two clear peaks at ~1568 cm^{-1} and ~2680 cm^{-1} which indicates that graphene is a good quality monolayer.

A gold contact (anode) was evaporated on the sample through a shadow mask. The other contact (cathode) was formed by coating silver paste on the scratched back-side of the Si substrate. The I-V measurements were performed with a Keithley Semiconductor Characterization System 4200 (SCS-4200) connected to a Janis probe station. During the measurements the sample was kept in dark and at a pressure of 1 mbar.

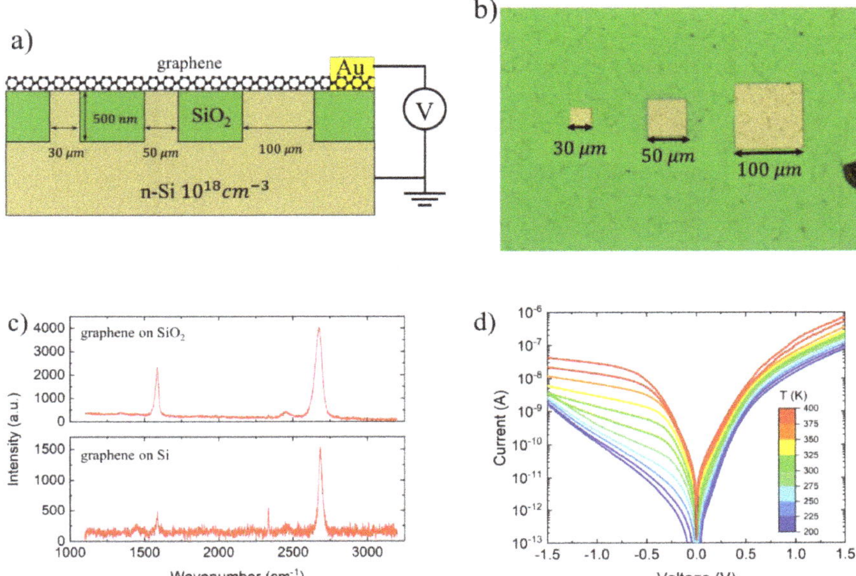

Figure 1. (**a**) Two-dimensional (2D) schematic view of the gr/Si-pillar device. (**b**) Optical microscope image of the pillars. (**c**) Raman spectroscopy of the graphene on SiO_2 and Si. (**d**) The current-voltage (I-V) characteristic of the device measured from 200 K to 400 K.

3. Results

Figure 1d shows the I-V characteristics measured for the gr/Si-pillar junction at different temperatures in the range 200–400 K. From low to room temperature the gr/Si junction shows an exponential reverse current which is typical of gr/Si junctions [17]. At higher temperatures, after the initial fast growth of the ohmic regime at low bias, the reverse current exhibits a gradual weaker dependence on the bias until it becomes quasi-saturated. The I-V characteristic at room temperature shows a rectification factor of two orders of magnitude at ±1.5 V.

The exponential reverse current growth at lower temperatures in Figure 1d can be explained considering the Fermi level shift due to the graphene low density of states, which reduces the Schottky barrier in reverse bias [31]. The variation of the barrier can be contributed also by the geometry and doping level of the substrate through the image-force barrier lowering. The pillar geometry magnifies the electric field around the perimeter where a wider depletion layer is created. Such a depletion layer is mirrored by charges in graphene, which cause an up-shift of the Fermi level and a reduction of the Schottky barrier. The high doping of the Si substrate can further contribute to barrier lowering through the image force effect. Conversely, the change of behaviour at higher temperatures indicates that the augmenting thermal generation rate in the depletion layer dominates the reverse leakage current which becomes less sensitive to the bias. The slight deviation of such current from saturation can be ascribed to image force barrier lowering [38,39].

To determine the Schottky diode parameters, we use the T.E. model with voltage dependent Schottky barrier height $q\phi_B$ [31], expressed by the equations:

$$I = I_0 e^{\frac{qV}{nkT}}\left(1 - e^{-\frac{qV}{kT}}\right), \quad (1)$$

$$I_0 = AA^* T^2 e^{-\frac{q\phi_B}{kT}}, \quad (2)$$

where I_0 is the reverse saturation current, q the electron charge, $n > 1$ the ideality factor, k the Boltzmann constant, T the temperature, A the junction area, $A^* = \frac{4\pi m_e^* k^2}{h^3} = 112\ \text{A cm}^{-2}\ \text{K}^{-2}$ the Richardson constant for n-type Si (m_e^* is the electron effective mass and h is the Plank constant) [40]. For $qV > nkT$, Equations (1) and (2) can be rewritten as:

$$\ln(I) = \ln(I_0) + \frac{qV}{nkT}, \tag{3}$$

$$\ln\left(\frac{I_0}{T^2}\right) = \ln(AA^*) - \frac{q\phi_B}{kT}. \tag{4}$$

According to Equation (3), the straight-line fitting of the ln(I)-V characteristics for $qV >> kT$ can be used to extrapolate the reverse current I_0 at zero bias and to estimate the ideality factor n. The so-obtained ideality factor as a function of temperature is shown in Figure 2a. The ideality factor at room temperature is $n \approx 2.6$ and is a monotonic decreasing function of the temperature because several non-idealities manifest more at lower temperatures. These non-idealities include metal residues consequence of the etching process (Cu in this case) which form carrier recombination centers, interface states at the junction which lead to charge trapping and detrapping, and the presence of a native oxide layer [31,34]. The zero-bias current, I_0, is used in the Richardson plot, $\ln\left(I_0/T^2\right)$ vs $1/T$, shown in Figure 2b, which, according to Equation (4), yields a Schottky barrier at zero-bias of 0.11 eV and $\ln(AA^*) = -33.72$. Since the effective gr/Si junction contact area is $\sim 1.34 \cdot 10^{-2}\ \text{mm}^2$, the Richardson constant is $A^* = 1.68{\cdot}10^{-9}\ \text{Acm}^2\text{K}^{-2}$. A possible explanation for the low Richardson constant and the ideality factor $n > 2$ is the presence of a thin oxide layer [16]. Taking into account the native oxide thickness, Equation (2) can be modified by adding a tunnelling factor as:

$$I_0 = AA^* \exp\left(-\chi^{\frac{1}{2}}\delta\right)\exp-\left(\frac{q\phi_B}{kT}\right), \tag{5}$$

where δ (expressed in Å) is the oxide layer thickness and $\chi \approx 3$ eV is the differences between the energy Fermi level and the conduction band minimum of SiO_2. From Equation (5), we estimated an oxide layer of 15 Å, which is thin enough to allow a tunnelling current, but can sustain a voltage drop and affect the I-V characteristic of the junction.

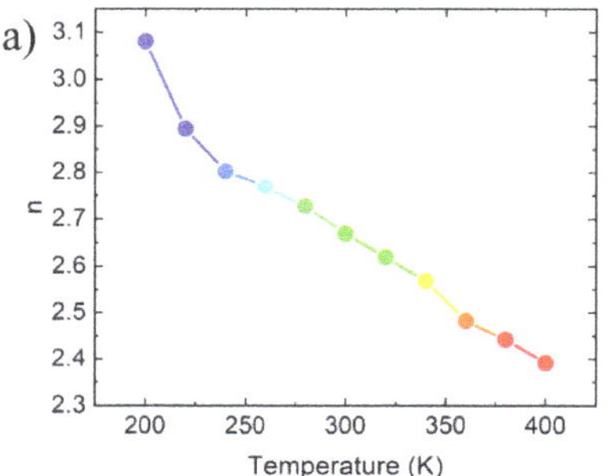

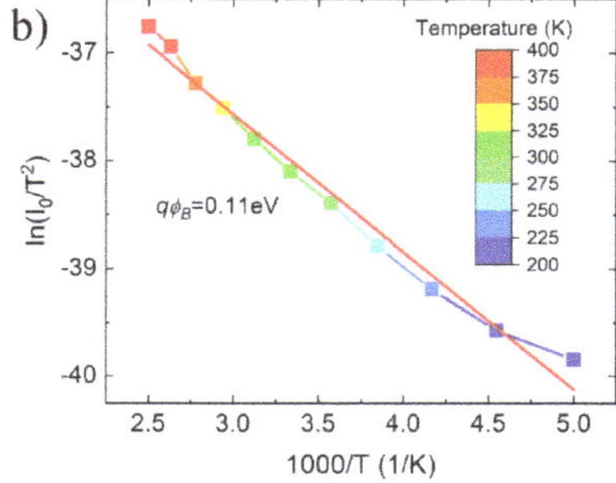

Figure 2. (**a**) Ideality factor vs the temperature extracted from the thermionic emission (T.E.) model (**b**) Richardson plot of the $\ln\left(I_0/T^2\right)$ versus $10^3/T$.

At higher positive bias ($V \gtrsim 0.8\ V$), the thermionic emission current is limited by the series resistance R_S, which is the lump sum of contact, graphene and substrate resistances. By taking it into account, Equation (1) can be rewritten as

$$I = I_0 e^{\frac{q(V-IR_S)}{nkT}}, \tag{6}$$

And from Equation (6), two new equations can be derived when $V - IR_S > nkT/q$ [37]:

$$\frac{dV}{d(\ln(i))} = IR_S - \frac{nkT}{q}, \tag{7}$$

$$H(I) = IR_S + n\phi_B, \tag{8}$$

where $H(I)$ is defined as:

$$H(I) = V - \frac{nkT}{q}\ln\left(\frac{I}{AA^*T^2}\right). \tag{9}$$

Accordingly, the series resistance and the ideality factor can be extrapolated from the slope and the intercept of the $dVd(\ln(I))$ vs I plot (Figure 3a), respectively, while the Schottky barrier can be estimated from the intercept of $H(I)$ vs I plot (Figure 3b). Using this method, at room temperature, we obtain 10 MΩ series resistance and ideality factor ~3. Figure 3c,d display the series resistance, the ideality factor and $q\phi_B$ measured at different temperatures. The decreasing series resistance with increasing temperature shows the typical semiconductor behaviour. This behaviour cannot be attributed to silicon, Au or Ag paste in this temperature range [41–43]. Therefore, it can only be caused by the graphene layer. The resistance drop at high temperature and the negative dR_S/dT has been reported for both exfoliated and CVD grown graphene [44–46]. The graphene semimetal behaviour has been attributed mainly to the thermally activated transport through the inhomogeneous electron-hole puddles, the formation of which is favoured by the transfer process of CVD-grown graphene [35,46].

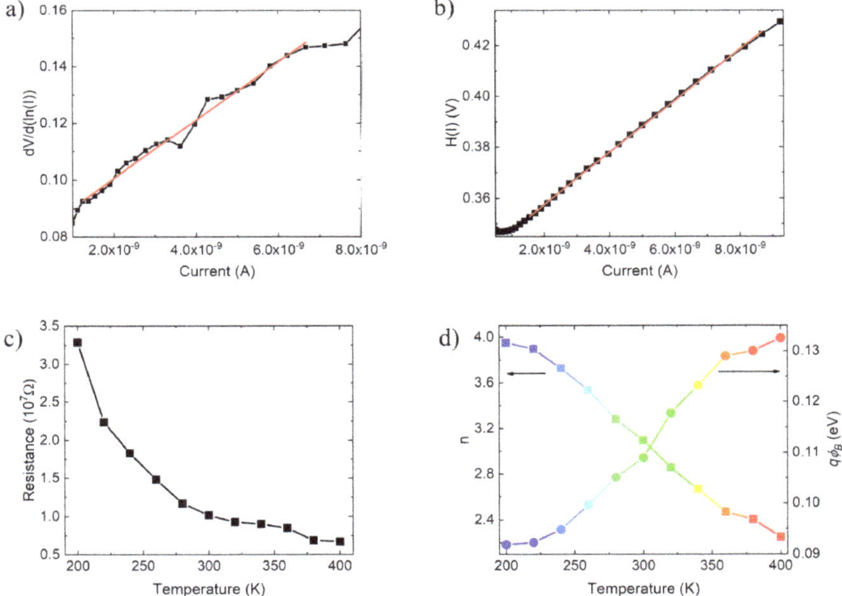

Figure 3. Cheung's plot of (**a**) $dV/d\ln(I)$ vs I and (**b**) $H(I)$ vs I at 300K. (**c**) Devices series resistance, (**d**) ideality factor and the Schottky barrier extracted from the Cheung and Cheung (CC) method versus the temperature.

Using Equations (7)–(9), we estimate $q\phi_B$ at different temperatures (Figure 3d); in particular $q\phi_B \approx 0.11$ eV at room temperature which is in agreement with the previous evaluation. The temperature growing $q\phi_B$ is an indication of possible spatial inhomogeneities. The homogeneity of the barrier will be discussed later. In Figure 4a,b we show the Richardson plot at given forward and reverse biases. In forward bias, the temperature dependence of the current has a linear behaviour, which is in agreement with the T.E. theory. Contrarily, in reverse bias, the evolving behaviour of the current, from exponential to saturation trend, is reflected in the Richardson plot (Figure 4b), which for $T \leq 300$ K is similar to the forward bias one (Figure 4a), while at higher temperature shows rising converging curves. Because of this, we consider only the lower temperature part of the curves in Figure 4b ($T \leq 300$ K) to determinate the Schottky barrier and the $\ln(AA^*)$, which are displayed in Figure 4c. We highlight that the Schottky barrier increases with the applied voltage, as expected. In forward bias, the graphene Fermi energy shifts down with respect to the semiconductor energy bands, thus increasing the Schottky barrier, while the opposite occurs in reverse bias. The relative shift, and therefore the barrier variation, is enhanced by the magnified electric field of the pillar and is made possible by the depinning of the Fermi level caused by the thin interfacial oxide layer [17,40].

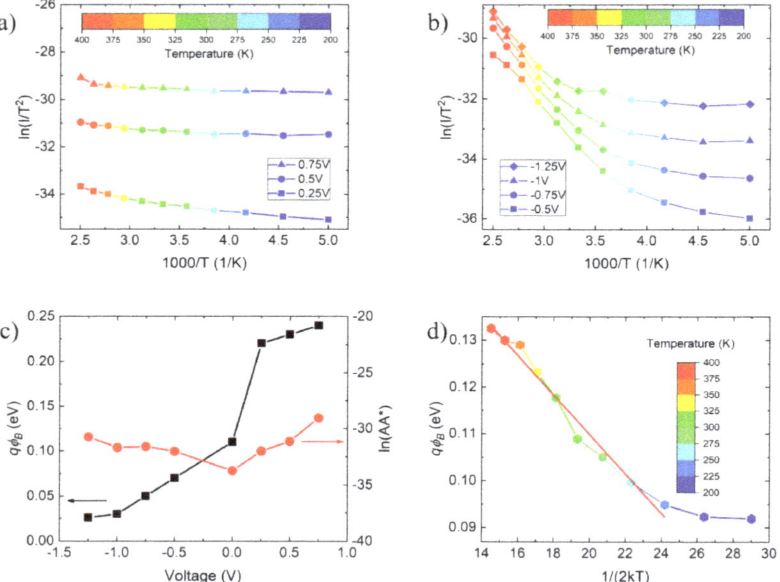

Figure 4. (**a**) Richardson plot of $\ln\left(I/T^2\right)$ vs 10^3 T in forward and (**b**) in reverse bias. (**c**) Schottky barrier and $\ln(AA^*)$ respect the bias. (**d**) Schottky barrier height at zero bias as a function of temperature.

Because of the CMP treatment (see Section 2), there is a possibility that the pillar top surface is not homogeneous and there could be points where the Schottky barrier is higher or lower. Following Refs. [17,40,47], we assume that the spatial variation of the Schottky barrier can be described by a Gaussian distribution. Therefore, the temperature dependence of the barrier is expressed as:

$$q\phi_B = q\phi_{BM} - \frac{q\sigma^2}{2kT}, \quad (10)$$

where $q\phi_{BM}$ is the maximum Schottky barrier and σ is the standard deviation of the Gaussian distribution. σ characterizes the inhomogeneity of the Schottky barrier and can be extracted from a plot of $q\phi_B$ vs $1/2kT$ (Figure 4d). We obtain $\sigma = 45$ meV, which is lower than those reported in literature for CVD grown graphene [48,49]. Since the graphene was CVD grown, the low standard variation can be considered as a remarkable advantage of the patterning of the substrate.

Finally, we measured the gr/Si response to light. Figure 5a shows the semi-logarithmic I-V curves of the device measured under different white LED light intensities. The responsivity

$\mathcal{R} = (I_{light} - I_{dark})/P_{opt}$ (I_{light} and I_{dark} are the current measured at $-1V$ under illumination and in dark, respectively) as a function of the incident light power P_{opt} is shown in Figure 5b. The device presents a responsivity with a maximum of ~88 A/W at 10^{-5}–10^{-4} Wcm^{-2}, which decreases at higher intensities. The reduction of the responsivity at higher intensities is due to the raising recombination rate. Indeed, at high illumination, the increasing of electron-hole pair density in the depletion layer enhances the recombination rate thus making the photocurrent deviate from its linearly behaviour as shown Figure 5c.

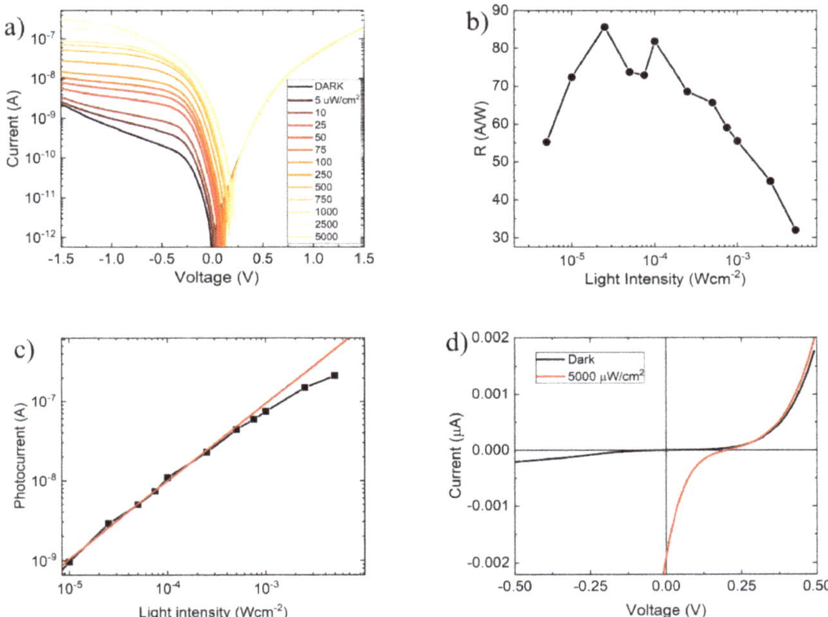

Figure 5. (**a**) I-V characteristic in semilogarithmic scale of the gr/Si pillar device measured at different intensity illumination level. (**b**) Responsivity of the gr/Si pillar device as function of the light intensity. (**c**) Photocurrent measured at $-1V$ and at different light intensities in logarithmic scale. (**d**) I-V characteristic measured in dark (black line) and at 5 mWcm^{-2} (red line).

Remarkably, the device achieves a reverse current that can be greater than the forward one. The high reverse current measured at high illumination confirms that there is a contribution to the junction current from the photogeneration occurring in the substrate areas where graphene forms a MOS capacitor with Si, as explained in previous works [16,18,36,50]. Furthermore, we note that the photogeneration has the same effect as the thermal generation in shaping the I-V curves of the device. Figure 5d shows the I-V measured in dark and under illumination at 5 mW/cm^2 in linear scale. A photovoltaic effect with an open circuit voltage around 0.19 V, which is close to the estimated Schottky barrier height, and a short circuit current of 1.8 nA, corresponding to ~0.7% power conversion efficiency, can be clearly observed. The conversion efficiency can be improved by lowering the doping of the Si substrate, which would result in an extended depletion layer for enhanced light absorption, and by reducing the shunt and series resistance that would increase the cell fill factor.

4. Conclusions

In conclusion, we fabricated a gr/Si pillar junction that possesses both a bias-tunable Schottky barrier, remarkable photoresponse and appreciable power conversion efficiency. The reverse current grows exponentially with reverse bias at lower temperatures, while it shows a saturation at higher temperatures or under illumination. Such behaviour has been explained by taking into account the thermo- and photo- generated charges both at the gr/Si junction and in the surrounding regions.

Author Contributions: A.D.B. conceived the experiment, G.L., A.G., M.L. and C.A.C. performed the experiment, G.L., L.I., F.G. and A.D.B. analysed the data, A.D.B., F.G. and C.W. contributed reagents/materials/analysis tools, G.L., F.G. and A.D.B. wrote the article.

Funding: We acknowledge the economic support of POR Campania FSE 2014–2020, Asse III Ob. specifico l4, D.D. n. 80, 31/05/2016 and CNR-SPIN SEED Project 2017 and Project PICO & PRO, ARS S01_01061, PON "Ricerca e Innovazione" 2014–2020.

Acknowledgments: We thank Technology Department, IHP-Microelectronics, Frankfurt Oder, Germany for the fabrication of the devices.

Conflicts of Interest: The authors declare no conflict of interest.

References

1. Geim, A.K.; Novoselov, K.S. The rise of graphene. *Nat. Mater.* **2007**, *6*, 183–191. [CrossRef] [PubMed]
2. Ganatra, R.; Zhang, Q. Few-Layer MoS_2: A promising layered semiconductor. *ACS Nano* **2014**, *8*, 4074–4099. [CrossRef]
3. Di Bartolomeo, A.; Genovese, L.; Giubileo, F.; Iemmo, L.; Luongo, G.; Foller, T.; Schleberger, M. Hysteresis in the transfer characteristics of MoS_2 transistors. *2D Mater.* **2017**, *5*, 015014. [CrossRef]
4. Li, H.; Lu, G.; Wang, Y.; Yin, Z.; Cong, C.; He, Q.; Wang, L.; Ding, F.; Yu, T.; Zhang, H. Mechanical exfoliation and characterization of single- and few-layer nanosheets of WSe_2, TaS_2, and $TaSe_2$. *Small* **2013**, *9*, 1974–1981. [CrossRef]
5. Di Bartolomeo, A.; Urban, F.; Passacantando, M.; McEvoy, N.; Peters, L.; Iemmo, L.; Luongo, G.; Romeo, F.; Giubileo, F. A WSe_2 vertical field emission transistor. *Nanoscale* **2019**, *11*, 1538–1548. [CrossRef] [PubMed]
6. Carvalho, A.; Wang, M.; Zhu, X.; Rodin, A.S.; Su, H.; Castro Neto, A.H. Phosphorene: From theory to applications. *Nat. Rev. Mater.* **2016**, *1*, 16061. [CrossRef]
7. Castro, N.A.H.; Guinea, F.; Peres, N.M.R.; Novoselov, K.S.; Geim, A.K. The electronic properties of graphene. *Rev. Modern Phys.* **2009**, *81*, 109–162. [CrossRef]
8. Xia, F.; Perebeinos, V.; Lin, Y.; Wu, Y.; Avouris, P. The origins and limits of metal–graphene junction resistance. *Nat. Nanotechnol.* **2011**, *6*, 179–184. [CrossRef]
9. Di Bartolomeo, A.; Santandrea, S.; Giubileo, F.; Romeo, F.; Petrosino, M.; Citro, R.; Barbara, P.; Lupina, G.; Schroeder, T.; Rubino, A. Effect of back-gate on contact resistance and on channel conductance in graphene-based field-effect transistors. *Diam. Relat. Mater.* **2013**, *38*, 19–23. [CrossRef]
10. Bartolomeo, A.D.; Giubileo, F.; Romeo, F.; Sabatino, P.; Carapella, G.; Iemmo, L.; Schroeder, T.; Lupina, G. Graphene field effect transistors with niobium contacts and asymmetric transfer characteristics. *Nanotechnology* **2015**, *26*, 475202. [CrossRef] [PubMed]
11. Giubileo, F.; Di Bartolomeo, A. The role of contact resistance in graphene field-effect devices. *Prog. Surf. Sci.* **2017**, *92*, 143–175. [CrossRef]
12. Di Bartolomeo, A.; Giubileo, F.; Iemmo, L.; Romeo, F.; Russo, S.; Unal, S.; Passacantando, M.; Grossi, V.; Cucolo, A.M. Leakage and field emission in side-gate graphene field effect transistors. *Appl. Phys. Lett.* **2016**, *109*, 023510. [CrossRef]
13. Yang, H.; Heo, J.; Park, S.; Song, H.J.; Seo, D.H.; Byun, K.-E.; Kim, P.; Yoo, I.; Chung, H.-J.; Kim, K. Graphene barristor, a triode device with a gate-controlled schottky barrier. *Science* **2012**, *336*, 1140–1143. [CrossRef]
14. Mehr, W.; Dabrowski, J.; Scheytt, J.C.; Lippert, G.; Xie, Y.-H.; Lemme, M.C.; Ostling, M.; Lupina, G. Vertical graphene base transistor. *IEEE Electr. Dev. Lett.* **2012**, *33*, 691–693. [CrossRef]
15. Chen, C.-C.; Aykol, M.; Chang, C.-C.; Levi, A.F.J.; Cronin, S.B. Graphene-silicon schottky diodes. *Nano Lett.* **2011**, *11*, 1863–1867. [CrossRef]
16. Di Bartolomeo, A.; Luongo, G.; Giubileo, F.; Funicello, N.; Niu, G.; Schroeder, T.; Lisker, M.; Lupina, G. Hybrid graphene/silicon Schottky photodiode with intrinsic gating effect. *2D Mater.* **2017**, *4*, 025075. [CrossRef]
17. Di Bartolomeo, A.; Giubileo, F.; Luongo, G.; Iemmo, L.; Martucciello, N.; Niu, G.; Fraschke, M.; Skibitzki, O.; Schroeder, T.; Lupina, G. Tunable Schottky barrier and high responsivity in graphene/Si-nanotip optoelectronic device. *2D Mater.* **2016**, *4*, 015024. [CrossRef]

18. Luongo, G.; Giubileo, F.; Genovese, L.; Iemmo, L.; Martucciello, N.; Di Bartolomeo, A. I-V and C-V characterization of a high-responsivity graphene/silicon photodiode with embedded MOS capacitor. *Nanomaterials* **2017**, *7*, 158. [CrossRef]
19. Luongo, G.; Di Bartolomeo, A.; Giubileo, F.; Chavarin, C.A.; Wenger, C. Electronic properties of graphene/p-silicon Schottky junction. *J. Phys. D Appl. Phys.* **2018**, *51*, 255305. [CrossRef]
20. Luongo, G.; Giubileo, F.; Iemmo, L.; Di Bartolomeo, A. The role of the substrate in Graphene/Silicon photodiodes. *J. Physi. Conf. Ser.* **2018**, *956*, 012019. [CrossRef]
21. Kim, H.-Y.; Lee, K.; McEvoy, N.; Yim, C.; Duesberg, G.S. Chemically modulated graphene diodes. *Nano Lett.* **2013**, *13*, 2182–2188. [CrossRef]
22. Singh, A.; Uddin, M.A.; Sudarshan, T.; Koley, G. Tunable Reverse-biased graphene/silicon heterojunction schottky diode sensor. *Small* **2014**, *10*, 1555–1565. [CrossRef]
23. Fattah, A.; Khatami, S.; Mayorga-Martinez, C.C.; Medina-Sánchez, M.; Baptista-Pires, L.; Merkoçi, A. Graphene/silicon heterojunction schottky diode for vapors sensing using impedance spectroscopy. *Small* **2014**, *10*, 4193–4199. [CrossRef] [PubMed]
24. Riazimehr, S.; Bablich, A.; Schneider, D.; Kataria, S.; Passi, V.; Yim, C.; Duesberg, G.S.; Lemme, M.C. Spectral sensitivity of graphene/silicon heterojunction photodetectors. *Solid State Electron.* **2016**, *115*, 207–212. [CrossRef]
25. Bonaccorso, F.; Sun, Z.; Hasan, T.; Ferrari, A.C. Graphene photonics and optoelectronics. *Nat. Photonics* **2010**, *4*, 611–622. [CrossRef]
26. An, X.; Liu, F.; Jung, Y.J.; Kar, S. Tunable graphene–silicon heterojunctions for ultrasensitive photodetection. *Nano Lett.* **2013**, *13*, 909–916. [CrossRef] [PubMed]
27. Ferrari, A.C.; Bonaccorso, F.; Fal'ko, V.; Novoselov, K.S.; Roche, S.; Bøggild, P.; Borini, S.; Koppens, F.H.L.; Palermo, V.; Pugno, N.; et al. Science and technology roadmap for graphene, related two-dimensional crystals, and hybrid systems. *Nanoscale* **2015**, *7*, 4598–4810. [PubMed]
28. An, X.; Liu, F.; Kar, S. Optimizing performance parameters of graphene–silicon and thin transparent graphite–silicon heterojunction solar cells. *Carbon* **2013**, *57*, 329–337. [CrossRef]
29. Behura, S.K.; Nayak, S.; Mukhopadhyay, I.; Jani, O. Junction characteristics of chemically-derived graphene/p-Si heterojunction solar cell. *Carbon* **2014**, *67*, 766–774. [CrossRef]
30. Ruan, K.; Ding, K.; Wang, Y.; Diao, S.; Shao, Z.; Zhang, X.; Jie, J. Flexible graphene/silicon heterojunction solar cells. *J. Mater. Chem. A* **2015**, *3*, 14370–14377. [CrossRef]
31. Di Bartolomeo, A. Graphene Schottky diodes: An experimental review of the rectifying graphene/semiconductor heterojunction. *Phys. Rep.* **2016**, *606*, 1–58. [CrossRef]
32. Liang, S.-J.; Hu, W.; Di Bartolomeo, A.; Adam, S.; Ang, L.K. A modified Schottky model for graphene-semiconductor (3D/2D) contact: A combined theoretical and experimental study. In Proceedings of the 2016 IEEE International Electron Devices Meeting (IEDM), San Francisco, CA, USA, 3–7 December 2016; pp. 14.4.1–14.4.4.
33. Tung, R.T. Electron transport at metal-semiconductor interfaces: General theory. *Phys. Rev. B* **1992**, *45*, 13509–13523. [CrossRef]
34. Zhang, X.; Zhang, L.; Ahmed, Z.; Chan, M. Origin of nonideal graphene-silicon schottky junction. *IEEE Transact. Electr. Dev.* **2018**, *65*, 1995–2002. [CrossRef]
35. Lupina, G.; Kitzmann, J.; Costina, I.; Lukosius, M.; Wenger, C.; Wolff, A.; Vaziri, S.; Östling, M.; Pasternak, I.; Krajewska, A.; et al. Residual metallic contamination of transferred chemical vapor deposited graphene. *ACS Nano* **2015**, *9*, 4776–4785. [CrossRef] [PubMed]
36. Riazimehr, S.; Kataria, S.; Bornemann, R.; Haring Bolívar, P.; Ruiz, F.J.G.; Engström, O.; Godoy, A.; Lemme, M.C. High photocurrent in gated graphene–silicon hybrid photodiodes. *ACS Photonics* **2017**, *4*, 1506–1514. [CrossRef] [PubMed]
37. Cheung, S.K.; Cheung, N.W. Extraction of Schottky diode parameters from forward current-voltage characteristics. *Appl. Phys. Lett.* **1986**, *49*, 85–87. [CrossRef]
38. Zhang, X.; Zhang, L.; Chan, M. Doping enhanced barrier lowering in graphene-silicon junctions. *Appl. Phys. Lett.* **2016**, *108*, 263502. [CrossRef]
39. Di Bartolomeo, A.; Grillo, A.; Urban, F.; Iemmo, L.; Giubileo, F.; Luongo, G.; Amato, G.; Croin, L.; Sun, L.; Liang, S.-J.; et al. Asymmetric schottky contacts in bilayer MoS_2 field effect transistors. *Adv. Funct. Mater.* **2018**, *28*, 1800657. [CrossRef]

40. Sze, S.M.; Ng, K.K. *Physics of Semiconductor Devices*, 3rd ed.; Wiley-Interscience: Hoboken, NJ, USA, 2007; ISBN 978-0-471-14323-9.
41. Li, S.S.; Thurber, W.R. The dopant density and temperature dependence of electron mobility and resistivity in n-type silicon. *Solid State Electron.* **1977**, *20*, 609–616. [CrossRef]
42. Jacoboni, C.; Canali, C.; Ottaviani, G.; Alberigi Quaranta, A. A review of some charge transport properties of silicon. *Solid State Electron.* **1977**, *20*, 77–89. [CrossRef]
43. Swirhun, S.E.; Swanson, R.M. Temperature dependence of specific contact resistivity. *IEEE Electr. Dev. Lett.* **1986**, *7*, 155–157. [CrossRef]
44. Shao, Q.; Liu, G.; Teweldebrhan, D.; Balandin, A.A. High-temperature quenching of electrical resistance in graphene interconnects. *Appl. Phys. Lett.* **2008**, *92*, 202108. [CrossRef]
45. Skákalová, V.; Kaiser, A.B.; Yoo, J.S.; Obergfell, D.; Roth, S. Correlation between resistance fluctuations and temperature dependence of conductivity in graphene. *Phys. Rev. B* **2009**, *80*, 153404. [CrossRef]
46. Heo, J.; Chung, H.J.; Lee, S.-H.; Yang, H.; Seo, D.H.; Shin, J.K.; Chung, U.-I.; Seo, S.; Hwang, E.H.; Das Sarma, S. Nonmonotonic temperature dependent transport in graphene grown by chemical vapor deposition. *Phys. Rev. B* **2011**, *84*, 035421. [CrossRef]
47. Werner, J.H.; Güttler, H.H. Barrier inhomogeneities at Schottky contacts. *J. Appl. Phys.* **1991**, *69*, 1522–1533. [CrossRef]
48. Parui, S.; Ruiter, R.; Zomer, P.J.; Wojtaszek, M.; van Wees, B.J.; Banerjee, T. Temperature dependent transport characteristics of graphene/n-Si diodes. *J. Appl. Phys.* **2014**, *116*, 244505. [CrossRef]
49. Tomer, D.; Rajput, S.; Hudy, L.J.; Li, C.H.; Li, L. Inhomogeneity in barrier height at graphene/Si (GaAs) Schottky junctions. *Nanotechnology* **2015**, *26*, 215702. [CrossRef]
50. Di Bartolomeo, A.; Luongo, G.; Iemmo, L.; Urban, F.; Giubileo, F. Graphene–Silicon Schottky Diodes for Photodetection. *IEEE Transact. Nanotechnol.* **2018**, *17*, 1133–1137. [CrossRef]

Article

Visible Light Driven Photoanodes for Water Oxidation Based on Novel r-GO/β-$Cu_2V_2O_7$/TiO_2 Nanorods Composites

Shuang Shuang [1,2,3,†], Leonardo Girardi [1,†], Gian Andrea Rizzi [1,*], Andrea Sartorel [1], Carla Marega [1], Zhengjun Zhang [3] and Gaetano Granozzi [1]

1 University of Padova and INSTM Unit, via Marzolo 1, 35121 Padova, Italy; shuangshuang_buct@163.com (S.S.); leonardo.girardi@phd.unipd.it (L.G.); andrea.sartorel@unipd.it (A.S.); carla.marega@unipd.it (C.M.); gaetano.granozzi@unipd.it (G.G.)

2 State Key Laboratory of New Ceramics and Fine Processing, School of Materials Science and Engineering, Tsinghua University, Beijing 100084, China

3 Key Laboratory of Advanced Materials (MOE), School of Materials Science and Engineering, Tsinghua University, Beijing 100084, China; zjzhang@mail.tsinghua.edu.cn

* Correspondence: gianandrea.rizzi@unipd.it; Tel.: +39-049-827-5722

† These authors contributed equally to this work.

Received: 4 June 2018; Accepted: 16 July 2018; Published: 18 July 2018

Abstract: This paper describes the preparation and the photoelectrochemical performances of visible light driven photoanodes based on novel r-GO/β-$Cu_2V_2O_7$/TiO_2 nanorods/composites. β-$Cu_2V_2O_7$ was deposited on both fluorine doped tin oxide (FTO) and TiO_2 nanorods (NRs)/FTO by a fast and convenient Aerosol Assisted Spray Pyrolysis (AASP) procedure. Ethylenediamine (EN), ammonia and citric acid (CA) were tested as ligands for Cu^{2+} ions in the aerosol precursors solution. The best-performing deposits, in terms of photocurrent density, were obtained when NH_3 was used as ligand. When β-$Cu_2V_2O_7$ was deposited on the TiO_2 NRs a good improvement in the durability of the photoanode was obtained, compared with pure β-$Cu_2V_2O_7$ on FTO. A further remarkable improvement in durability and photocurrent density was obtained upon addition, by electrophoretic deposition, of reduced graphene oxide (r-GO) flakes on the β-$Cu_2V_2O_7$/TiO_2 composite material. The samples were characterized by X-ray Photoelectron Spectroscopy (XPS), Raman, High Resolution Transmission Electron Microscopy (HR-TEM), Scanning Electron Microscopy (SEM), Wide Angle X-ray Diffraction (WAXD) and UV–Vis spectroscopies. The photoelectrochemical (PEC) performances of β-$Cu_2V_2O_7$ on FTO, β-$Cu_2V_2O_7$/TiO_2 and r-GO/β-$Cu_2V_2O_7$/TiO_2 were tested in visible light by linear voltammetry and Electrochemical Impedance Spectroscopy (EIS) measurements.

Keywords: copper vanadate; photoanode; water splitting; graphene oxide

1. Introduction

The hydrogen economy is a new intriguing sustainable scenario, and it is expected that sooner or later it is going to replace the hydrocarbon economy [1]. With this perspective, a worldwide goal is to provide sustainable and convenient methods to prepare hydrogen fuel. Among them, water splitting (WS), by exploiting the energy of sun, is the most appealing, and many different approaches are currently being investigated [2]. The advantage of the photoelectrochemical (PEC) approach, compared to the standard photocatalytic one, is that an external potential is used to facilitate the WS process. The intrinsic simplicity of PEC, which combines the light absorber and the energy converter into a single device capable to store solar energy into chemical bonds, is rather evident. For these reasons, numerous studies have been performed to fabricate semiconducting nanomaterials with enhanced PEC properties under visible light. In particular, many efforts have been done on developing materials for photoanodes, where the kinetically hindered Oxygen Evolution Reaction (OER) is occurring. Their final performance depends upon the electrocatalysts stability against oxidation [3] and on their intrinsic band energetics [4].

Metal oxide semiconductors are promising photoanode materials because of their relative stability to oxidative photo-corrosion and their low-cost. Hematite, (α-Fe_2O_3), has been identified as an efficient photoanode material characterized by a sufficiently low band gap of 1.9–2.0 eV, to be used as "top" electrode in a "tandem" WS device [5]. Nevertheless, this material has some shortcomings, such as a short carrier diffusion length, a significant recombination and indirect absorption. During the last few years, other multicomponent oxides have been suggested as possible active materials for the construction of photoanodes. Among them, $ZnFe_2O_4$, $CuWO_4$ $CuW_{1-x}Mo_xO_4$ and especially Cu-vanadates are particularly studied. As with many transition metal (TM) vanadates, Cu-vanadates are characterized by different phases where the Cu/V ratio has a quite large variability. The performances of these phases, together with their PEC stability, were tested in a very comprehensive paper by Gregoire et al. [6]. In their study it was shown that sputter-deposited phases having lower Cu/V ratios are less stable, in borate buffer solution (pH = 9.2), than other phases with higher Cu/V ratio, (e.g., γ-$Cu_3V_2O_8$ and $Cu_{11}V_6O_{26}$). The γ-$Cu_3V_2O_8$ based photoanodes, prepared by sol-gel method were also recently studied by Neale et al. [7]. The results of these studies were that the V-rich phases suffered from V loss and a consequent decay in the PEC properties, while the higher stability of Cu rich phases was attributed to a self-passivating mechanism that led to the formation of Cu^+ and Cu^{2+} oxides on the vanadate surface. In a successive study, again by Gregoire et al. [4], a library of Cu-vanadates thin films with variable stoichiometry was prepared by a fast and convenient ink-jet printing procedure and, again, it resulted that both α-CuV_2O_6 and α-$Cu_2V_2O_7$ are highly active and stable photocatalysts in a borate buffer solution, while β-$Cu_2V_2O_7$ demonstrated a high photoelectroactivity in the presence of ferri/ferrocyanide redox couple at pH = 13. The abovementioned phases, α-CuV_2O_6 and β-$Cu_2V_2O_7$ were also deposited by a simple drop casting method on fluorine doped tin oxide (FTO) glass and their PEC properties characterized by Mullins et al. [8]. This study showed that the V rich α-CuV_2O_6 phase is the one showing the highest photocurrent, although both phases were characterized by a short diffusion length for holes and required the addition of a hole scavenger like Na_2SO_3 to improve the photocurrent density. Finally, in a more recent paper by Sharp et al. [9] it was shown that, although Cu-rich phases show higher absorption and charge separation, these phases also present a higher surface recombination rate. Therefore, considering that Cu-rich phases are the ones showing the higher stability in borate buffer solutions, good charge separation and higher absorption and that, on the other hand, V rich phases seem to be those characterized by the higher photocurrents density, we decided to concentrate our attention on the β-$Cu_2V_2O_7$ phase, that appeared to be a good compromise between durability and PEC performances. The idea of increasing the adhesion between the substrate and the vanadate particles, in order to improve the durability in the electrolyte solution, led us to think about the use of a high surface area substrate like TiO_2 nanorods (NRs) on FTO [10], to grow this n-type semiconductor. Moreover, the addition of graphene oxide (GO) flakes, could lead to the formation of a composite material with interesting PEC proprieties in term of durability and photocurrent density. In fact, quite recently, composite systems like $BiVO_4/TiO_2$ and $V_2O_5/BiVO_4/TiO_2$ were prepared by hydrothermal synthesis and, although not used as active material in photoanodes, showed superior photocatalytic performances in the degradation of organics caused by an upward shift of V_2O_5 and $BiVO_4$ conduction bands with respect to TiO_2 with formation of an n-n junction [11,12]. A similar approach was also used by Chen and coworkers [13] where TiO_2 NRs were decorated by Fe_2O_3 grown on preformed TiO_2 NRs obtained by a simple hydrothermal synthesis. With respect to the effect of the addition of GO flakes, it is useful to remind that GO and especially partial reduced GO (r-GO), are considered as a good support for nanostructures because of their carrier mobility [14], large specific area and high optical transmittance [15,16]. Moreover, when TiO_2 nanostructures are combined with GO or r-GO they usually can shuttle and store more electrons due to the formation of many p-n nanojunctions with r-GO, a p-type semiconductor [17–19].

In this study, we describe the preparation and evaluation of the PEC performances of a visible light driven photoanode based on a novel composite material consisting on β-$Cu_2V_2O_7$ nanoparticles deposited on TiO_2 NRs followed by the addition of r-GO flakes. The decoration of TiO_2 NRs with β-$Cu_2V_2O_7$ NPs was obtained by an easy and fast aerosol assisted spray pyrolysis (AASP) deposition technique. A further improvement of the performances was obtained by the addition of partially reduced graphene oxide (r-GO), a p-type semiconductor [20], to reduce the charge transfer resistance. We show here that GO, deposited by electrophoretic deposition, can efficiently coat the surface of the β-$Cu_2V_2O_7$/TiO_2 nanostructures and, after a mild annealing, is transformed into r-GO causing a remarkable enhancement of the photocurrent with increased durability (from 50 $\mu A/cm^2$ in the case of pure TiO_2 NRs, 150 $\mu A/cm^2$ for TiO_2NRs decorated with β-$Cu_2V_2O_7$ NP to 250 $\mu A/cm^2$ for r-GO/β-$Cu_2V_2O_7$/TiO_2).

The prepared films were characterized by wide angle X-ray diffraction (WAXD), scanning electron microscopy (SEM), high resolution transmission electron microscopy (HR-TEM), UV–Vis and Raman spectroscopy, Electrochemical Impedance Spectroscopy (EIS) and PEC measurements. Their surface composition was also studied by X-ray Photoemission Spectroscopy (XPS) before and after the PEC work, under illumination, with the intention of verifying V loss and concomitant formation of CuO_x passivating layers.

2. Materials and Methods

2.1. Material Preparation

All the reagents used in this study were analytical grade and purchased from Sigma-Aldrich (Milan, Italy). The photoactive material of this study, β-$Cu_2V_2O_7$ was deposited on FTO or FTO/TiO_2 substrates by a quick and convenient AASP method. This method consisted of the evaporation of a precursor solution micro-droplets onto a heated substrate. The micro-droplets were produced by the nebulization created through ultrasounds. Using a stream of gas (i.e., air), the micro droplets were transported near the substrate. Malachite [$Cu_2(OH)_2CO_3$] was used as a copper-source for the aerosol solution because it contains "clean" anionic groups (carbonate and hydroxide) that do not introduce other contamination into the solution. For the preparation of $Cu_2(OH)_2CO_3$, potassium hydrogen carbonate (10.0 g, 100 mmol) and copper sulphate pentahydrate (10.0 g, 40 mmol) were dissolved in 150 mL of hot water within two beakers. The two solutions were mixed together after being cooled. Immediately, a teal blue precipitate formed. These precipitates were recovered by "Büchner" filtration and washed with water and ethanol. The reaction that takes place is:

$$2CuSO_4(aq) + 4KHCO_3(aq) + H_2O(l) \rightarrow Cu_2(OH)_2CO_3(s) + 2K_2SO_4(aq) + 3CO_2\ (g)$$

The solid was dried on a hot plate at ca. 200 °C; this was because at a temperatures close to 300 °C, malachite starts to decompose [21]. The aerosol precursor solution was prepared using a malachite suspension in water followed by the addition of a suitable ligand to complex Cu^{2+} ions, thus preventing the direct precipitation of copper vanadate, once the vanadate source is added. After complete dissolution of the malachite suspension, the vanadium-source (NH_4VO_3) was added. Ammonia, citric acid (CA) and ethylenediamine (EN) were used as ligands to prepare precursor solutions 1, 2 and 3, respectively.

Solution 1 $Cu_2(OH)_2CO_3$ (0.23 g, 1 mmol) was added to 10 mL of deionized water followed by the addition of 2 mL of concentrated (33%) NH_3 solution, under vigorous stirring. Ammonium vanadate, NH_4VO_3 (0.25 g, 2 mmol), was added until a clear solution was formed.

Solution 2 CA (0.63 g, 3 mmol) and malachite (0.23 g, 1 mmol) were dissolved in 10 mL of deionized water. Ammonium vanadate (0.25 g, 2 mmol) was added to this solution.

Solution 3 EN (150 μL, 2 mmol) and, malachite (0.23 g, 1 mmol) were dissolved in 10 mL of deionized water. Ammonium vanadate (0.25 g, 2 mmol) was added to this solution.

TiO_2 NRs were fabricated on FTO glass (TiO_2/FTO) by the hydrothermal method [22]. 2 FTO slides (1 cm × 1.5 cm × 0.2 cm) were cleaned with isopropanol and deionized water in a sonicator for 30 min. The precursor solution was prepared by combining 3.44 mL of deionized water with 3.44 mL of HCl 37% and mixing for 10 min. Finally, 120 μL of titanium isopropoxide [$Ti(OCH(CH_3)_2)_4$] were added under vigorous stirring. This solution was poured into the Teflon-liner with the FTO substrates together. The whole system was heated to 150 °C for 4 h and cooled down to room temperature (RT), after extraction from the oven. The FTO slides were finally rinsed with abundant deionized water.

Deposition of copper vanadate, either on clean FTO or on TiO_2 NRs, was carried out using a commercial aerosol medical device (Artsana Projet). To avoid ammonia evaporation, in the case of solution 1, an excess of ammonia water solution was used (4 mL of solution 1 and 2 mL of 33% ammonia water solution).

The aerosol was conveyed through a tube (flow rate of about 60 mL/min) to a funnel neck, just above the substrate that was positioned on a metal plate heated by a Boraelectric heater (Tectra, GmbH, Frankfurt, Germany), connected to a power supply. A K-type thermocouple was positioned between the plate and the heater, to have accurate control of the sample temperature. The substrate was heated for 20 min to 340 °C. During the deposition the temperature decreased to 320 °C. The optimal deposition time on FTO was found to be 5 min, while on TiO_2/FTO, the best results were obtained after 3 min. The TiO_2/FTO and the β-$Cu_2V_2O_7$/FTO samples were then annealed in air at 450 °C for 2 and 4 h, respectively.

GO was synthetized from graphite using an Improved Hummers' method [23], developed from Marcano et al.; GO water suspension (2.5 mg/mL; pH = 6.5) was prepared using a sonicator to disperse the flakes.

The deposition was carried out by an electrophoretic process (2.5 mg/mL GO water suspension, pH = 6.5). A 5V potential was applied for 30 s between the sample (positive pole) and a clean FTO glass. The FTO slides were separated by a distance of 1.5 cm. After the deposition, the sample was annealed in air at 200 °C for 15 min.

2.2. Material Characterization

The morphology and nanostructure of all samples were characterized by field-emission gun SEM (Zeiss Supra 35VP, Zeiss, Jena, Germany) and High-Resolution TEM (JEOL-2011, JEOL Ltd., Tokyo, Japan). Surface composition was determined by XPS measurements performed on a custom-built UHV chamber (base pressure = 5×10^{-10} mbar) equipped with a non-monochromatized double-anode X-ray source (Omicron DAR-400, Scienta-Omicron GmbH, Uppsala, Sweden), a hemispherical electron analyzer (Omicron EA-125, Scienta-Omicron GmbH, Uppsala, Sweden) and a 5-channeltrons detection assembly. The electron analyzer had an acceptance angle of ±4° and the diameter of the analyzed area was 3 mm. The spectra were acquired with Al-Kα radiation. WAXD patterns were recorded in the diffraction angular range 10−50° 2θ by a Philips X'Pert PRO diffractometer, working in the reflection geometry and equipped with a graphite monochromator on the diffracted beam (CuKα radiation, Pananlytical, Almelo, The Netherlands). Raman spectra were acquired with a Thermo-Fisher DXR Raman microscope using a 532 nm laser (5 mW), focused on the sample with a 50× objective (Thermo-Fisher Scientific, Madison, WI, USA) obtaining a spot size of about 1 μm. UV–Vis spectra were acquired in absorbance and reflectance mode on a UV–Vis Cary 5E spectrophotometer.

All the electrochemical measurements were obtained in a Na borate buffer solution prepared adding NaOH to a 0.4 M solution of boric acid until pH = 9.2 was reached (example of PEC measurement obtained in Na-sulphate solution reported in Figures S1 and S2). The measurements were made in a Teflon PEC cell (see Figure S3). A Pt wire and Ag/AgCl electrode were used as counter electrode and reference electrode, respectively. PEC measurements were obtained by a visible light emitting diode (LED) source (see Figure S4) controlled by the optical bench (Metrohm-Autolab) coupled to the Autolab PGSTAT204 (Metrohm, Utrecht, The Netherlands) instrument. The samples were mounted outside the cell and kept in position by an O-ring seal. All samples were illuminated from the back side (comparison between front-side and back-side illumination reported in Figure S5) and the electrical contact was obtained by a Cu strip attached to the FTO glass surface by Silver Conductive Paint (RS). EIS data were obtained under illumination and in the dark at 1.75 V vs. Reversible Hydrogen Electrode (RHE). The amplitude for EIS measurements was ±10 mV with the frequency range set from 10^5 to 10^{-1} Hz, performing 50 points with logarithmic distribution. Oxygen measurement was carried out by NEOFOX-KIT PROBE from Ocean Optics (Ocean Optics, 8060 Bryan Dairy Rd, Largo, FL 33777, USA).

3. Results and Discussion

The PEC measurements on β-$Cu_2V_2O_7$ films deposited on FTO (β-$Cu_2V_2O_7$) using different aerosol solutions (solution 1–3) are discussed later in the text, nevertheless, it is useful to anticipate that vanadates prepared using solution 1 (NH_3) gave the best PEC results with respect to the other two solutions. For this reason, only β-$Cu_2V_2O_7$ on TiO_2 NRs (β-$Cu_2V_2O_7$/TiO_2) and r-GO/β-$Cu_2V_2O_7$ on TiO_2 (r-GO/β-$Cu_2V_2O_7$/TiO_2) samples, obtained with solution 1 (see experimental section) are herein discussed. We attributed this behavior to a lower carbon contamination.

Figure 1a shows the Raman spectra of β-$Cu_2V_2O_7$, TiO_2, β-$Cu_2V_2O_7$/TiO_2 and r-GO/β-$Cu_2V_2O_7$/TiO_2, measured at room-temperature. The Raman region of pure TiO_2 NRs presents all the characteristic peaks corresponding to rutile, that is, the peak at 244 cm^{-1} corresponding to the phonon scattering mode of rutile, the signal at 438 cm^{-1} assigned to the E_g mode, and the peak at 621 cm^{-1} to the A_{1g} mode [24]. The peak centered at 914 cm^{-1} is the characteristic band assigned to the β-$Cu_2V_2O_7$ (VO_3 stretching mode) [25]. In the case of r-GO/ β-$Cu_2V_2O_7$/TiO_2 NRs sample, two broad peaks at 1354 (I_D) and 1598 cm^{-1} (I_G) are those characteristic of r-GO [26]. In particular, the I_D/I_G ratio corresponding to thick r-GO flakes was 0.97, as reported in reference [26], while in the case of areas where the r-GO coating was not visible by the micro-Raman microscope (50×), the ratio was 0.89 before PEC measurements and was reduced to 0.85 after PEC measurements (see Supplementary Material, Figure S6). These values are completely consistent with the presence of r-GO. Figure 1b shows the WAXD patterns of the prepared samples: β-$Cu_2V_2O_7$, β-$Cu_2V_2O_7$/TiO_2, and r-GO/β-$Cu_2V_2O_7$/TiO_2. The diffraction peak at 2θ = 24.7° (cyan curve) is assigned to reflections from planes (200) of monoclinic β-$Cu_2V_2O_7$ (JCPDS No. 73-1032), while peaks at 2θ = 36.2° and 62.9° (red, green and blue curves) correspond to reflections from planes (101) and (002) of rutile (JCPDS No. 21-1276). We calculated the lattice parameters of TiO_2-rutile NRs before and after the coating with $Cu_2V_2O_7$. These parameters are: a = 4.569(9) Å, c = 2.955(2) Å, remaining unchanged after the coating. The blue curve (r-GO/β-$Cu_2V_2O_7$/TiO_2) presents a further diffraction peak, at 2θ =24.7°, that has to be attributed to r-GO [27].

Indeed, the XRD pattern of β-$Cu_2V_2O_7$/TiO_2 (green curve) shows an extremely weak peak at 2θ = 24.7°. However, the very low intensity of this peak, probably due to the small thickness of the β-$Cu_2V_2O_7$ coating made it difficult to detect β-$Cu_2V_2O_7$ by X-ray diffraction and, therefore, the presence of the reflection at 2θ = 24.7°, in the case of the r-GO/β-$Cu_2V_2O_7$/TiO_2 has to be related to r-GO.

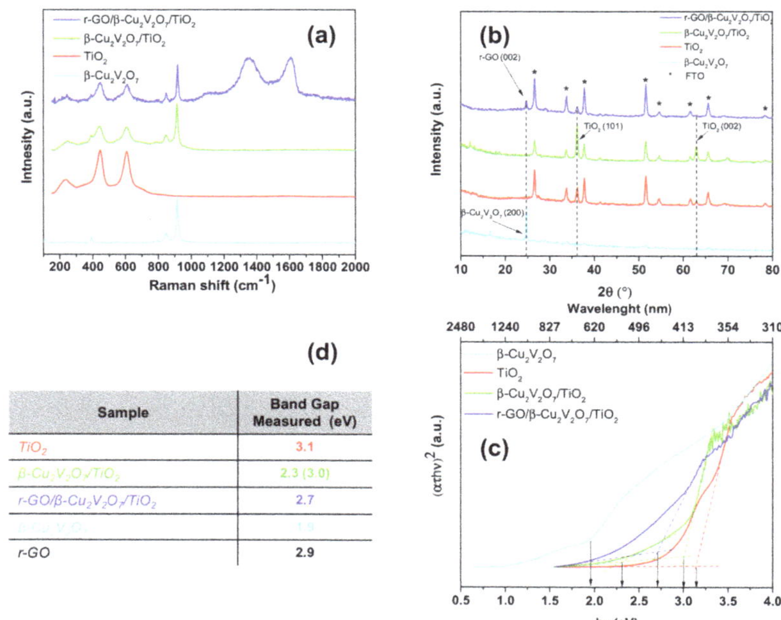

Sample	Band Gap Measured (eV)
TiO_2	3.1
β-$Cu_2V_2O_7/TiO_2$	2.3 (3.0)
r-GO/β-$Cu_2V_2O_7/TiO_2$	2.7
β-$Cu_2V_2O_7$	1.9
r-GO	2.9

Figure 1. Raman spectra (**a**); wide angle X-ray diffraction (WAXD) patterns (**b**); and Tauc plots (**c**) of β-$Cu_2V_2O_7$, TiO_2 NRs, β-$Cu_2V_2O_7/TiO_2$ and GO/β-$Cu_2V_2O_7/TiO_2$; Bandgap values obtained from Tauc plots are indicated in the table on the left (**d**).

The band gap values (E_g) of these semiconducting materials can be estimated from the Tauc plots (Figure 1c). The absorption coefficient is calculated from Equation (1):

$$C\alpha\tau = -\ln\left(\frac{T}{1 - R_{ref}}\right) \tag{1}$$

where α is the absorption coefficient, τ is the thickness of the film, C is a constant, T the transmittance, and R_{ref} the reflectance. Since all the samples studied were rather opaque it was necessary to acquire both the diffuse reflectance spectra and the transmittance spectra. The band gap (E_g) was estimated by calculating the intercept of an extrapolated linear fit of the experimental data, $[\alpha\tau h\nu]^2$, to the flat portion of the plot, where no absorption occurs. The measured values of E_g for a direct transition are shown in Figure 1d [28]. β-$Cu_2V_2O_7$ and TiO_2 show E_g values of 1.9 and 3.1 eV, respectively. Samples β-$Cu_2V_2O_7/TiO_2$ and r-GO/β-$Cu_2V_2O_7/TiO_2$ present a Tauc plot characterized by a shape typical of composite materials [11] with intercepts at ca. 3 eV (β-$Cu_2V_2O_7/TiO_2$ presents an additional band gap at 2.3 eV related to copper vanadate particles) and 2.7 eV, after addition of r-GO (blue curve).

Figure 2 shows the SEM images of the pure TiO_2 NRs supported on FTO (Figure 2a) and those decorated with β-$Cu_2V_2O_7$ (Figure 2c,d) and coated with r-GO flakes (Figure 2e,f). The as annealed film consists of TiO_2 NRs with a diameter of ~50 nm and a length of ~2 µm (Figure 2b). The sectional view, reported in Figure 2b, shows that these TiO_2 NRs are vertically aligned on the FTO substrate with a thickness of about 2–2.4 µm. After the deposition of β-$Cu_2V_2O_7$, the oxide nanoparticles stick randomly on the top of TiO_2 NRs surface (Figure 2c,d). Finally, the r-GO flakes tile the nanorods, similar to a silk coat. (Figure 2e,f).

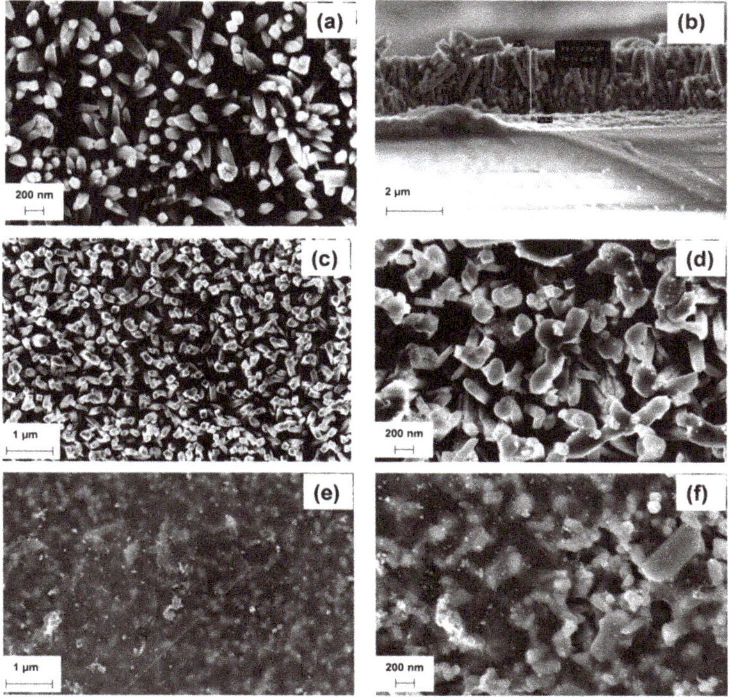

Figure 2. Scanning electron microscope (SEM) images of: TiO_2 nanorods (NRs) on fluorine doped tin oxide (FTO) (**a**); cross-section of TiO_2 NRs (**b**); β-$Cu_2V_2O_7/TiO_2$ (**c**,**d**); r-GO/β-$Cu_2V_2O_7/TiO_2$ (**e**,**f**).

The TEM images and the corresponding energy dispersive X-ray (EDX) images are also presented in Figure 3. The size of the β-$Cu_2V_2O_7$ NPs is between 100 and 200 nm with a regular cubic shape (Figure 3a–c). According to the measurement of the lattice fringes (d = 0.249, 0.320, and 0.307 nm) there is a very good match with the crystallographic planes of rutile (101), rutile (110) and β-$Cu_2V_2O_7$ (022), respectively (Figure 3d–f). The O, Ti, V and Cu EDX elemental maps are also reported in Figure 3h together with the physical images. These images show that V and Cu are not only present on the vanadate NPs, but also on the surface of the NRs. The AASP deposition procedure allows the deposition of β-$Cu_2V_2O_7$ crystals not only on top of the rods, but also along their length, with variable dimensions caused by the diffusion of aerosol droplets through the porous TiO_2 NRs layer.

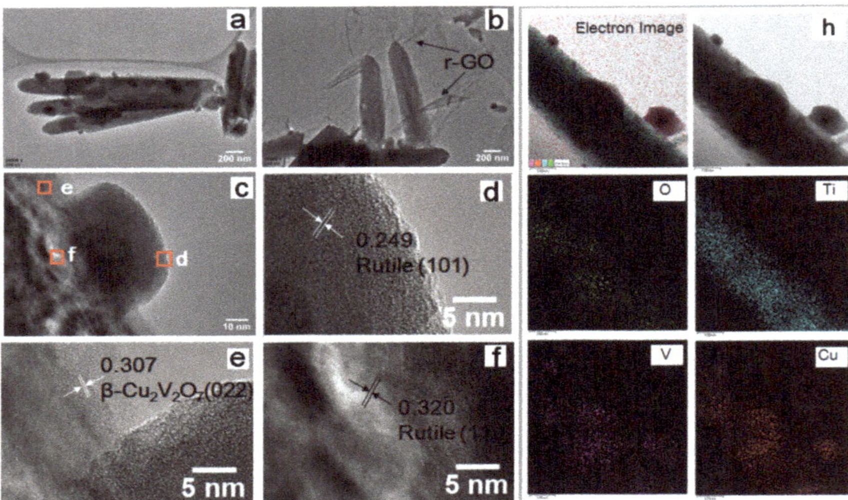

Figure 3. Transmission Electron Microscopy (TEM) images of β-$Cu_2V_2O_7/TiO_2$ (**a**); and r-GO/β-$Cu_2V_2O_7/TiO_2$ (**b**); High Resolution Transmission Electron Microscopy (HR-TEM) images of GO/β-$Cu_2V_2O_7/TiO_2$ (**c**–**f**); morphology and energy dispersive X-ray (EDX) elemental mapping of r-GO/β-$Cu_2V_2O_7/TiO_2$ NRs sample (**h**).

To obtain further information on the surface composition of these nanostructures the samples were characterized by XPS, before PEC measurements, as reported in Figure 4. Figure 4 shows the O 1s, V $2p_{3/2}$ and Cu $2p_{3/2}$ XPS spectra obtained from β-$Cu_2V_2O_7/TiO_2$ (Figure 4a–c) and r-GO/β-$Cu_2V_2O_7/TiO_2$ (Figure 4d–f). The O 1s XPS spectrum of β-$Cu_2V_2O_7/TiO_2$ (Figure 4a) can be fitted with two components, located at about 529.8 and 532.0 eV, corresponding to lattice O^{2-} ions from metal oxides and hydroxyl groups. In the case of the sample decorated with r-GO, the O 1s signal is mainly due to the oxygen atoms bound to carbon (Figure 4f) and can be fitted with three components at 531.0, 533.0 and 534.5 eV. These three components are due to (O=C) groups, alcoholic groups (HO–C) and water, respectively. The signal at about 529.9 eV, assigned to TiO_2 and vanadate lattice oxygens, is highly attenuated by the GO layers that coat the TiO_2 NRs (see SEM images) [29]. In the case of copper vanadate supported on TiO_2 NRs, without r-GO, the V $2p_{3/2}$ signal (Figure 4b) can be fitted with only one component at 516.8 eV with a full width at half maximum (FWHM) of about 1.5 eV, corresponding to V^{5+}, while in the case of the sample decorated with r-GO, Figure 4e, the signal contains two components at 516.4 and 517.5 eV corresponding to V^{4+} and V^{5+}, respectively [6]. It is interesting to note that the Cu $2p_{3/2}$ signal (Figure 4c,d) indicates the presence of Cu^{+2}, assigned to the component at 535.4 eV, and Cu^{+} at 533.0 eV [30]. The Cu^{+} signal, in the case of the sample treated with r-GO, is actually the main component (Cu^{2+} 41% and Cu^{+} 59%), indicating that some reaction has occurred between β-$Cu_2V_2O_7$ and GO. This is confirmed also by the presence of a quite high amount of V^{4+} (V^{5+} is 62% and V^{4+} is 38%) signal, while the Cu^{2+}/V^{5+} ratio (58% Cu^{2+} and 42% V^{5+}) is not too far from the 1:1 expected value for β-$Cu_2V_2O_7$. In the case of the β-$Cu_2V_2O_7/TiO_2$ sample, the obtained Cu^{2+}/V^{5+} ratio is also close to the expected value (40% of Cu and 60% of V) and the presence of Cu^{+} (Cu^{+} 35.5%, Cu^{2+} 64.5%) can be due to a photoreduction effect due to the X-ray source or to the presence of traces of CuO_x [6].

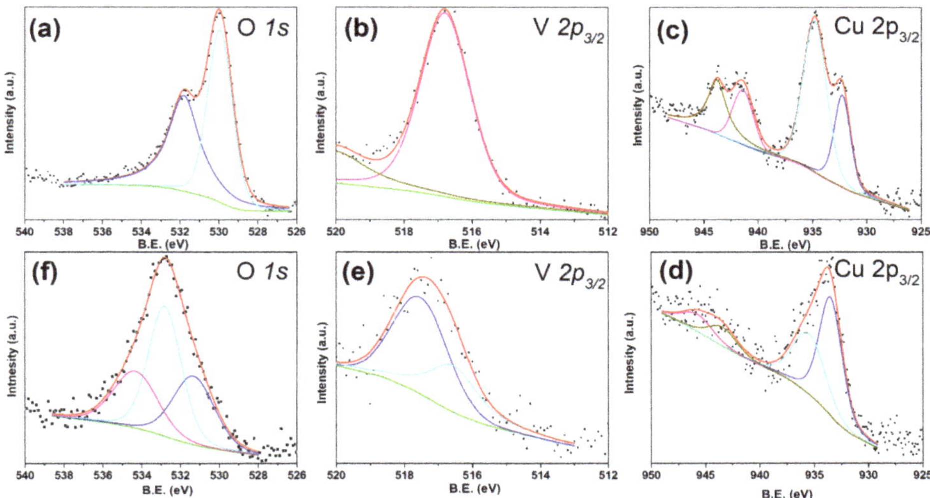

Figure 4. Spectroscopy (XPS) spectra of as-prepared β-$Cu_2V_2O_7/TiO_2$ (**a**–**c**); and r-GO/β-$Cu_2V_2O_7/TiO_2$ (**d**–**f**) samples.

All substrates were tested in PEC experiments, where the light source was a neutral white led with intensity ca. 100 mW/cm^2 (Figure S4) in Na-borate buffer electrolyte (pH = 9.2). In Figure 5a, we report the linear voltammetry scans under chopped light for pure β-$Cu_2V_2O_7$ deposits obtained using different ligands. From the plot it is easy see that NH_3 furnishes the better results in terms of photocurrent (ca. 220 μA/cm^2 a 1.55 V vs. RHE). For this reason, the decoration of TiO_2 NRs by β-$Cu_2V_2O_7$ was obtained by using NH_3 in the precursor solution. As clearly visible in Figure 5b,c, the TiO_2 NRs decorated with β-$Cu_2V_2O_7$ show a better performance in terms of durability with almost no variation in the photocurrent after 3 h of EC work. On the contrary, the photocurrent density is lower with respect to the pure, β-$Cu_2V_2O_7$ on FTO (Figure 5a).

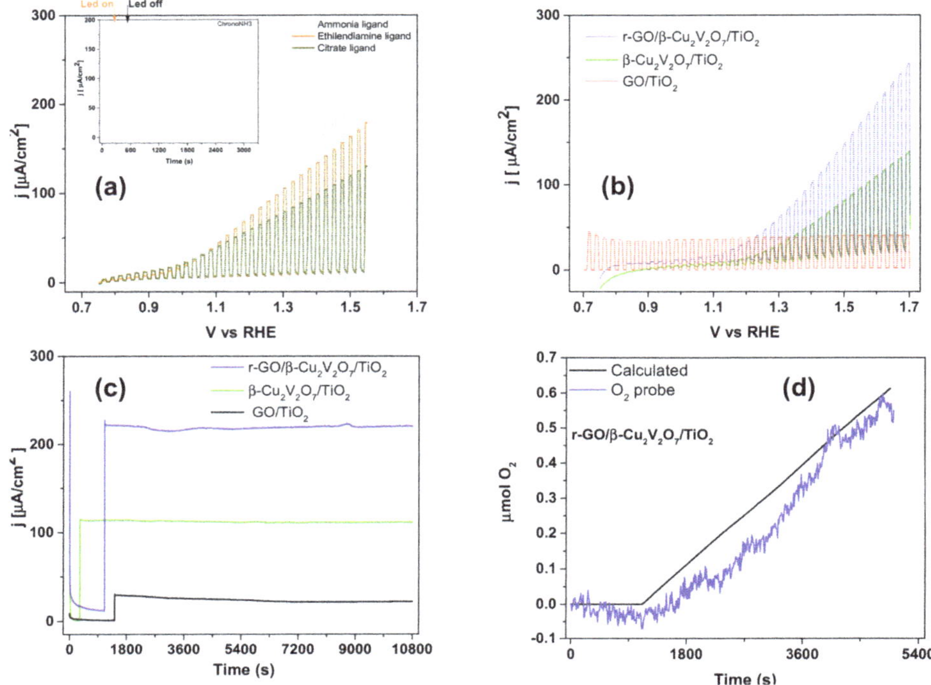

Figure 5. Photoelectrochemical performances: chopped Linear Sweep Voltammetry (LSV) (Borate Buffer pH = 9.2, scan rate 5 mV/s) of β-$Cu_2V_2O_7$ deposited from aerosol solutions containing NH_3, EN or CA, as ligands, on FTO (**a**) (the inset shows a chronoamperometry at 1.5 V vs. Reversible Hydrogen Electrode (RHE) of a sample deposited with ammonia as ligand); chopped LSV (Borate Buffer pH = 9.2, scan rate 5 mV/s) of samples deposited on TiO2 NRs (**b**); chronoamperometry of β-$Cu_2V_2O_7$/TiO_2, r-GO/β-$Cu_2V_2O_7$/TiO_2 and r-GO/TiO_2 at 1.5 V vs. RHE (**c**); comparison of calculated and measured O_2 when r-GO/β-$Cu_2V_2O_7$/TiO_2 is used as working electrode with light is set on at ca. 1000 s (**d**).

Addition of GO flakes by electrophoretic deposition allowed to obtain a much higher photocurrent density (see Figure 5b,c) and a very good durability. Finally, in Figure 5d, we report a comparison between the O_2 measured for the r-GO/β-$Cu_2V_2O_7$/TiO_2 sample in the gas phase (head-space in a sealed electrochemical cell previously purged with N_2), by an O_2 probe, based on the quenching of fluorescence, and the theoretical one, calculated from the recorded photocurrent. This measurement clearly demonstrates that the recorded photocurrent is not due to side processes like r-GO oxidation. The samples were also characterized by impedance spectroscopy (EIS) in the dark and under illumination at 1.5 V vs. RHE. From the data reported in Figure 6a,b it is evident that the samples β-$Cu_2V_2O_7$/TiO_2 and, especially r-GO/β-$Cu_2V_2O_7$/TiO_2, are characterized by a much lower charge transfer resistance. The equivalent circuit used to fit the data [31] contains 2 RQ elements (parallel connection of an ohmic resistance R and a constant phase element Q), in the case of pure β-$Cu_2V_2O_7$ on FTO, while for the composite materials β-$Cu_2V_2O_7$/TiO_2 and r-GO/β-$Cu_2V_2O_7$/TiO_2, we have used a series of 3 RQ elements. This circuit is represented in Figure 6d where the R_s represents the solution resistance, the first RQ element the double layer, the second one the Cu-vanadate layer and the third one the TiO_2 NRs. It is interesting to point out that upon illumination only the second circuit (R_2) shows a very strong decrease in the charge transfer resistance, while the other 2 circuits present only minor variations. This is a strong indication that it is mainly the Cu-vanadate layer that acts as the active material in the water photo-oxidation, while the role of TiO_2 is simply that of a substrate.

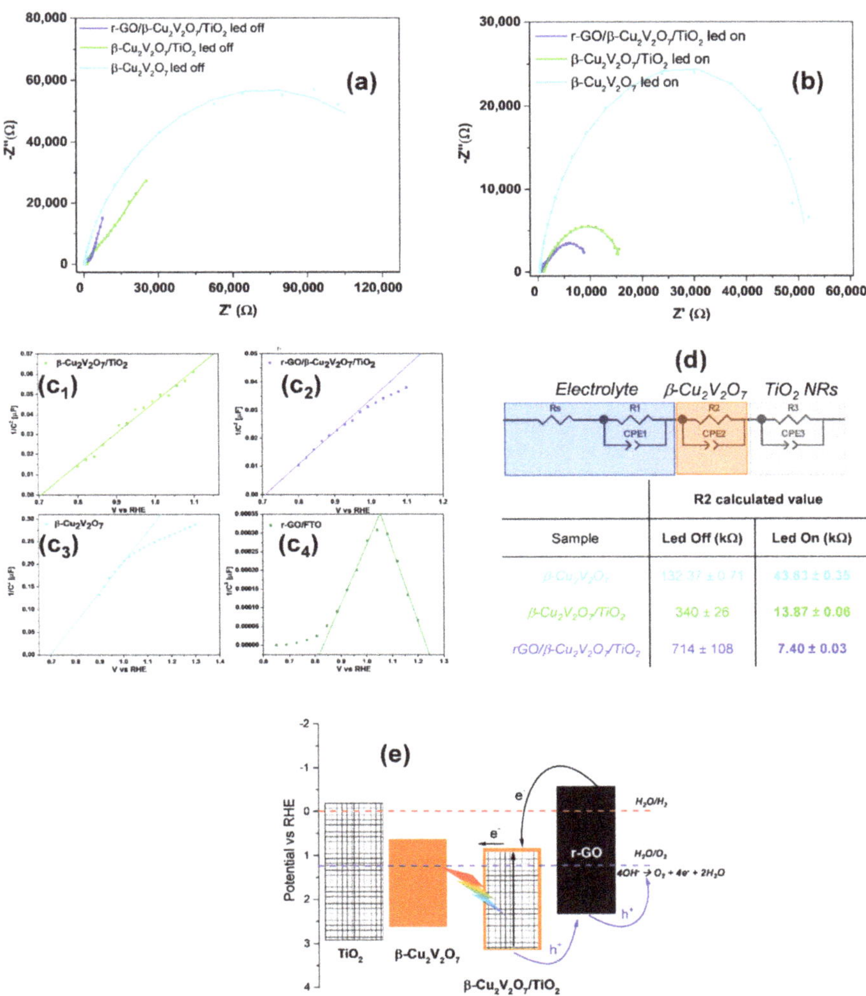

	R2 calculated value	
Sample	Led Off (kΩ)	Led On (kΩ)
β-$Cu_2V_2O_7$	132.37 ± 0.71	43.83 ± 0.35
β-$Cu_2V_2O_7/TiO_2$	340 ± 26	13.87 ± 0.06
rGO/β-$Cu_2V_2O_7/TiO_2$	714 ± 108	7.40 ± 0.03

Figure 6. Nyquist plots obtained with samples polarized at 1.5 V vs RHE in the dark (**a**); and under illumination (**b**); Mott-Schottky plots for β-$Cu_2V_2O_7/TiO_2$ (**c$_1$**), r-GO/ β-$Cu_2V_2O_7/TiO_2$ (**c$_2$**), β-$Cu_2V_2O_7$ (**c$_3$**), r-GO/FTO (**c$_4$**); EIS equivalent circuit [31] (**d**); schematic representation of band edges approximate position for r-GO/β-$Cu_2V_2O_7/TiO_2$ sample (**e**) Band edges positions for β-$Cu_2V_2O_7$ and TiO_2 are added for comparison.

In Figure 6c we report also the Mott-Schottky (MS) plots (in the range 1–10^5 Hz) obtained from pure β-$Cu_2V_2O_7$, β-$Cu_2V_2O_7/TiO_2$ and r-GO/β-$Cu_2V_2O_7/TiO_2$. The relation between the flat-band potential (V_{fb}) and the material capacity (C) is reported in Equation (2). N_{SC} indicates the carrier's concentration in the space charge of the material, ε the dielectric constant, e is the electron charge and

A is the area of the electrode. n-types semiconductors, like TiO_2 and β-$Cu_2V_2O_7$, are characterized by positive slopes, while p-types materials have negative slopes.

$$\frac{1}{C^2} = \frac{2\left(V - V_{fb}\right)}{eN_{sc}\varepsilon A^2} \quad (2)$$

The capacity values were calculated by fitting the impedance data with a Randle's circuit containing Constant Phase Elements (CPE) instead of ideal capacitors. Thus, the capacity was calculated from Brugg's Equation (3) [32]

$$C = (Q)^{\frac{1}{p}}\left(R_S^{-1} + R_p^{-1}\right)^{(1-\frac{1}{p})} \quad (3)$$

where Q and p, are fitting parameters from CPE elements, R_s is the cell resistance and R_p is the resistance in parallel with CPE elements. By plotting C^{-2} vs. RHE it is possible to determine V_{fb} and from this value to derive the approximate position of conduction (CB) and valence (VB) edges. The relation between V_{fb} and bands edges (E_{cb} and E_{vb}) can be expressed by Equations (4) and (5) [33]:

$$E_{cb} = V_{fb} + k_b T \ln\left(\frac{N_{sc}}{N_{cb}}\right) \quad (4)$$

$$E_{vb} = -V_{fb} + k_b T \ln\left(\frac{N_{sc}}{N_{vb}}\right) \quad (5)$$

where N_{cb} and N_{vb}, are the effective density of states in the CB and VB for a n-type and p-type semiconductors, respectively. In the case of n-type conductivity Equation (4) is usually approximated with $E_{cb} \approx V_{fb}$ + 0.1 eV [8]. Thus, the MS plots reported in Figure 6c, show how the decoration of TiO_2 does not change band edges position of the copper vanadate (TiO_2 acts as a support), while the addition of p-type GO, probably produces many p-n nano-junctions with β-$Cu_2V_2O_7/TiO_2$ (see scheme of Figure 6d, although not visible from the MS plot of Figure 6c_2. Indeed, the lower slope of the MS plot of Figure 6c_2 indicates a higher concentration of electrons (1×10^{17} m^{-3} for β-$Cu_2V_2O_7/TiO_2$ and 1.4×10^{17} m^{-3} for the sample decorated with r-GO), as already seen in the case of TiO_2 nanorods decorated with Cu_2O nanoparticles [34]. Finally, the p-type conductivity of r-GO is clearly seen from the MS plot obtained from a pure r-GO sample deposited on FTO and thermally treated at 200 °C for 15 min (Figure 6c_4).

More precise details about the surface composition of these nanostructures can be obtained by acquiring XPS data after electrochemical work. The results of this analysis are summarized in Figure 7 and Table 1. Figure 7a,b shows the O 1s, V 2p and Cu 2p XPS spectra obtained from β-$Cu_2V_2O_7/TiO_2$ NRs after 3 h of electrochemical measurements, under illumination. The O 1s XPS spectrum can be fitted with two components located at about 529.9 and 532.0 eV that correspond to lattice O^{2-} ions from β-$Cu_2V_2O_7$ and TiO_2 and hydroxyls groups [35]. The V $2p_{3/2}$ signal (in Figure 7a), can be fitted with two components at 516.0, weak, and 517.0 eV corresponding to V^{4+} and V^{5+} respectively. The Cu $2p_{3/2}$ signals (Figure 7b) contains two components, one at 935.0 and another at 933.0 eV indicating the presence of Cu^{2+} and a considerable quantity of Cu^{+}. The presence of Cu^{+} is probably caused by a photo-reduction effect and eventually by X-rays in UHV. In Figure 7c we show also the C 1s spectra acquired from a sample of r-GO/β-$Cu_2V_2O_7/TiO_2$ after PEC work. The region can be fitted with 3 components at 284.1, 285.7 and 288.0 eV corresponding respectively to C–C, C–O and C=O bonds [26]. The relative intensity and positions of these signals are fully compatible with p-type r-GO oxide, after a mild heat treatment [36]. A simple visual inspection of the O 1s and V 2p region, reported in Figure 7d, reveals how the amount of V in the case of the r-GO/β-$Cu_2V_2O_7/TiO_2$ is significantly lower if compared with the sample not containing GO. In fact, the V 2p signal is much lower with respect to the O 1s signal at 529.4 eV. Moreover, the Cu 2p signal, reported in Figure 7e, shows mostly the presence of Cu^{+} deduced from the position (933.0 eV) and the very low intensity of the satellites peaks. This fact is in agreement with what already verified on the sample before PEC work where the high amount of Cu^{+} and the high Cu/V ratio indicated that the addition of GO modified the composition of the vanadate.

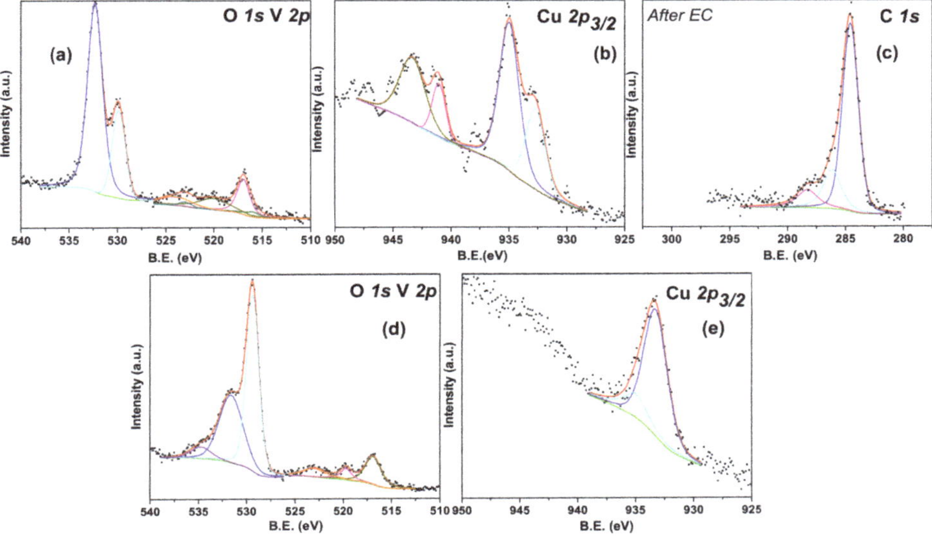

Figure 7. XPS spectra of samples β-$Cu_2V_2O_7/TiO_2$ after Electrochemical work (**a**–**b**); and of r-GO/β-$Cu_2V_2O_7/TiO_2$ (**c**–**e**) after photoelectrochemical (PEC) work.

The Cu/V ratios obtained from XPS data after EC work are similar to those obtained before EC (Raman spectra and SEM images after EC are reported in Figures S6 and S7, respectively). In the case of β-$Cu_2V_2O_7/TiO_2$ we found a 52% abundance of V^{5+} and 48% for Cu^{2+}, with a rather high amount of Cu^{+} (Figure 7b). In the case of the sample decorated with r-GO, the amount of Cu^{2+} is 32%, while the amount of V^{5+} is 68%. In this last case the large quantity of Cu^{+} seems to be due to the presence of the GO layer.

Table 1. Cu and V percent abundance from XPS data.

Element Abundance	Before PEC		After PEC	
Sample	Cu^{2+}	V^{5+}	Cu^{2+}	V^{5+}
r-GO/β-$Cu_2V_2O_7/TiO_2$	58	42	32	68
β-$Cu_2V_2O_7/TiO_2$	40	60	52	48

All the above reported data indicate that supporting β-$Cu_2V_2O_7$ on TiO_2 NRs allows the obtainment of a material with good durability, as photoanode, with a much lower charge transfer resistance as testified by the EIS data (Figure 6a,b). The interaction between the TiO_2 NRs and β-$Cu_2V_2O_7$ has a favorable effect on the photocurrent production since the deposition of V_2O_5 on TiO_2 NRs, by the same AASP process, does not lead to any particular enhancement in the photocurrent if compared with untreated TiO_2 (see Figures S8 and S9). It's also known that the decoration of Degussa P25 by CuO nanoparticles actually leads to a modest enhancement of photocurrent values [37]. A significant improvement in the photocurrent density can be achieved by decorating the β-$Cu_2V_2O_7/TiO_2$ sample by GO followed by heat treatment at about 200 °C. The addition of GO changes the Cu/V ratio leading to high amount of Cu^{+} as testified by the XPS spectra, acquired before and after EC work. We think that the better performance in terms of photocurrent is due to the combination of r-GO/β-$Cu_2V_2O_7/TiO_2$ since the addition of r-GO to TiO_2 NRs did not lead to any particular improvement in the photocurrent, as is clearly visible in Figure 5b. We verified that the interaction of GO with β-$Cu_2V_2O_7$ and heat treatment at 200 °C, after the electrophoretic deposition of GO, actually leads to the formation of p-type r-GO with a band gap, as measured from the UV–Vis spectra (see Figure S10) of about 2.5–2.7 eV. We can justify the interaction of GO with β-$Cu_2V_2O_7$, that leads to the variation in the Cu/V ratio, as caused by the relatively low pH of the GO suspension (pH = 6.5), used during the electrophoretic deposition process, and by the GO itself. In fact, the solubility of β-$Cu_2V_2O_7$ increases at low pH and, at the same time, the GO sheets can easily co-ordinate the Cu^{2+} ions [38]. The Cu^{2+} ions, once chemisorbed on the GO nano-sheets, most probably by the carboxylic groups, can be reduced by GO and most probably by the heat treatment, with formation of CO_2 as summarized in the following reaction sequence [38]:

$$[Cu(H_2O)_4]^{2+} + GO \rightarrow GO\text{-}[Cu(H_2O)_4]^{2+}$$

$$GO\text{-}[Cu(H_2O)_4]^{2+} \rightarrow GO\text{-}[Cu(H_2O)_2]^{2+}$$

$$GO\text{-}[Cu(H_2O)_2]^{2+} + heat \rightarrow r\text{-}GO\text{-}[Cu(H_2O)_2]^{+} + CO_2$$

As known and reported in several publications, the mild thermal treatment at 200–210 °C that leads to the formation of a partially reduced GO has to be intended formally as a disproportionate reaction where the electrons released with O_2, CO or CO_2 evolution are used to reduce the GO surface [39,40]. In this particular case, we think that this process leads also to the formation of Cu^+ species as Cu_2O nanoparticles. Since the V_{fb} is the same for β-$Cu_2V_2O_7$, β-$Cu_2V_2O_7/TiO_2$ and r-GO/β-$Cu_2V_2O_7/TiO_2$ and the Cu^{2+}/V^{5+} ratio on r-GO/β-$Cu_2V_2O_7/TiO_2$ is compatible with a 1:2 value, we can formulate the hypothesis that the vanadate partially decomposes forming Cu_2O nanoparticles on the GO flakes and on the NRs surface, while remaining still on the TiO_2 surface, in lower amounts, in a form compatible with the a "CuV_2O_6" stoichiometry. The position of band edges of r-GO with respect to the band edges of the composite material, β-$Cu_2V_2O_7/TiO_2$, is particularly favorable to form many p-n nano-junctions, as depicted in Figure 6d, leading to a better charge separation, increase in the photocurrent density and improved durability. It is important to note that the slightly higher position of E_{cb} of pure TiO_2 NRs (slightly above the H^+/H_2 reduction potential) would have led to a less favorable junction with p-type r-GO. Finally, the possible formation of a further p-n junction between the Cu_2O nanoparticle and the TiO_2 surface should also be taken into account [34].

4. Conclusions

We have prepared β-$Cu_2V_2O_7/TiO_2$ and r-GO/β-$Cu_2V_2O_7/TiO_2$ composite materials with the aim of obtaining visible light driven photoanodes. The vanadate deposition was obtained by an easy & fast aerosol assisted spray pyrolysis procedure. The β-$Cu_2V_2O_7/TiO_2$ composite material showed a better durability if compared with pure β-$Cu_2V_2O_7$ deposited on FTO and a lower charge transfer resistance as indicated by EIS data. The addition of p-type r-GO to β-$Cu_2V_2O_7/TiO_2$ had a positive effect on durability, charge transfer resistance and photocurrent density. As verified by XPS analysis, the GO addition, by electrophoretic deposition, led to a strong interaction with β-$Cu_2V_2O_7$ with formation of r-GO flakes and Cu_2O nanoparticles. The amount of O_2 produced, upon visible light illumination, independently measured by an O_2 probe, indicated that this composite material is characterized by a good faradaic efficiency. The easy and fast procedure that allows its preparation can be easily extended, with low cost, to electrodes with a larger area.

Supplementary Materials: The following are available online at http://www.mdpi.com/2079-4991/8/7/544/s1. Figure S1. Chopped Linear Sweep Voltammetry of sample r-GO/β-$Cu_2V_2O_7/TiO_2$ in Na sulfate electrolyte; Figure S2. Chronoamperometry from sample r-GO/β-$Cu_2V_2O_7/TiO_2$ at 1.75 V vs. RHE; Figure S3. Drawing of the photoelectrochemical cell (Proteus Gamma I—PINE Research) (CE = Counter Electrode; RE = Reference

Electrode); Figure S4. Emission spectrum of white LED light used in all the reported PEC measurements; Figure S5. Chopped LSV on r-GO/β-$Cu_2V_2O_7$/TiO_2 sample with front and back illumination (ca. 100 mW/cm^2) in borate buffer solution; Figure S6. (a) Raman spectra before and after PEC work obtained from different areas of sample r-GO/β-$Cu_2V_2O_7$ after (b) and before (c) PEC work; Figure S7. SEM image of sample r-GO/β-$Cu_2V_2O_7$/TiO_2 after PEC work; Figure S8. Chopped Linear Sweep Voltammetry of the TiO_2 nanorods substrate in borate buffer (pH = 9.2) with led light intensity of ca. 100 mW/cm^2; Figure S9. Chopped Linear Sweep Voltammetry of TiO_2 NRs decorated with V_2O_5 nanoparticles in borate buffer (pH = 9.2); Figure S10. Tauc Plot of r-GO deposited by electrophoresis on FTO slides.

Author Contributions: G.A.R. conceived and designed the experiments; L.G., S.S., C.M. and A.S. performed the experiments; Z.Z. and G.A.R. analyzed the data; G.G. and G.A.R. wrote the paper. We would like to thank Luca Bardini for helpful discussions.

Acknowledgments: Authors gratefully acknowledge the Italian Minister of University (MIUR) for financial support to the SMARTNESS (Solar driven chemistry: new materials for photo- and electrocatalysis) financed thorough the PRIN 2015K7FZLH. Thanks are due to the MAECI (Ministero degli Affari Esteri e della Cooperazione Internazionale) for the support through the bilateral Italy−China GRAPE-MAT project. SS thanks the China Scholarship Council (CSC) for a grant to spend one year at University of Padova.

Conflicts of Interest: The authors declare no conflict of interest.

References

1. Muradov, N.Z.; Veziroğlu, T.N. From hydrocarbon to hydrogen-carbon to hydrogen economy. *Int. J. Hydrogen Energy* **2005**, *30*, 225–237. [CrossRef]
2. Walter, M.G.; Warren, E.L.; McKone, J.R.; Boettcher, S.W.; Mi, Q.; Santori, E.A.; Lewis, N.S. Solar water splitting cells. *Chem. Rev.* **2010**, *110*, 6446–6473. [CrossRef] [PubMed]
3. Osterloh, F.E. Inorganic Materials as Catalysts for Photoelectrochemical Splitting of Water. *Chem. Mater.* **2008**, *20*, 35. [CrossRef]
4. Newhouse, P.F.; Boyd, D.A.; Shinde, A.; Guevarra, D.; Zhou, L.; Soedarmadji, E.; Li, G.; Neaton, J.B.; Gregoire, J.M. Solar fuel photoanodes prepared by inkjet printing of copper vanadates. *J. Mater. Chem. A* **2016**, *4*, 7483–7494. [CrossRef]
5. Fountaine, K.T.; Lewerenz, H.J.; Atwater, H.A. Efficiency limits for photoelectrochemical water-splitting. *Nat. Commun.* **2016**, *7*, 1–9. [CrossRef] [PubMed]
6. Zhou, L.; Yan, Q.; Yu, J.; Jones, R.J.R.; Becerra-Stasiewicz, N.; Suram, S.K.; Shinde, A.; Guevarra, D.; Neaton, J.B.; Persson, K.A.; et al. Stability and self-passivation of copper vanadate photoanodes under chemical, electrochemical, and photoelectrochemical operation. *Phys. Chem. Chem. Phys.* **2016**, *18*, 9349–9352. [CrossRef] [PubMed]
7. Seabold, J.A.; Neale, N.R. All first row transition metal oxide photoanode for water splitting based on $Cu_3V_2O_8$. *Chem. Mater.* **2015**, *27*, 1005–1013. [CrossRef]
8. Guo, W.; Chemelewski, W.D.; Mabayoje, O.; Xiao, P.; Zhang, Y.; Mullins, C.B. Synthesis and Characterization of CuV_2O_6 and $Cu_2V_2O_7$: Two Photoanode Candidates for Photoelectrochemical Water Oxidation. *J. Phys. Chem. C* **2015**, *119*, 27220–27227. [CrossRef]
9. Jiang, C.M.; Segev, G.; Hess, L.H.; Liu, G.; Zaborski, G.; Toma, F.M.; Cooper, J.K.; Sharp, I.D. Composition-Dependent Functionality of Copper Vanadate Photoanodes. *ACS Appl. Mater. Interfaces* **2018**, *10*, 10627–10633. [CrossRef]
10. Liu, B.; Aydil, E.S. Growth of oriented single-crystalline rutile TiO_2 nanorods on transparent conducting substrates for dye-sensitized solar cells. *J. Am. Chem. Soc.* **2009**, *131*, 3985–3990. [CrossRef] [PubMed]
11. Sun, J.; Li, X.; Zhao, Q.; Ke, J.; Zhang, D. Novel V_2O_5/$BiVO_4$/TiO_2 nanocomposites with high visible-light-induced photocatalytic activity for the degradation of toluene. *J. Phys. Chem. C* **2014**, *118*, 10113–10121. [CrossRef]
12. Hu, Y.; Li, D.; Zheng, Y.; Chen, W.; He, Y.; Shao, Y.; Fu, X.; Xiao, G. $BiVO_4$/TiO_2 nanocrystalline heterostructure: A wide spectrum responsive photocatalyst towards the highly efficient decomposition of gaseous benzene. *Appl. Catal. B Environ.* **2011**, *104*, 30–36. [CrossRef]
13. Wang, C.; Chen, Z.; Jin, H.; Cao, C.; Li, J.; Mi, Z. Enhancing visible-light photoelectrochemical water splitting through transition-metal doped TiO_2 nanorod arrays. *J. Mater. Chem. A* **2014**, *2*, 17820–17827. [CrossRef]

14. Chen, X. juan; Dai, Y. zhi; Wang, X. yan; Guo, J.; Liu, T. hua; Li, F.F. Synthesis and characterization of Ag_3PO_4 immobilized with graphene oxide (GO) for enhanced photocatalytic activity and stability over 2,4-dichlorophenol under visible light irradiation. *J. Hazard. Mater.* **2015**, *292*, 9–18. [CrossRef] [PubMed]
15. Bai, Y.-Y.; Wang, F.-R.; Liu, J.-K. A New Complementary Catalyst and Catalytic Mechanism: Ag_2MoO_4 /Ag/AgBr/GO Heterostructure. *Ind. Eng. Chem. Res.* **2016**, *55*, 9873–9879. [CrossRef]
16. Thomas, A.V.; Andow, B.C.; Suresh, S.; Eksik, O.; Yin, J.; Dyson, A.H.; Koratkar, N. Controlled crumpling of graphene oxide films for tunable optical transmittance. *Adv. Mater.* **2015**, *27*, 3256–3265. [CrossRef] [PubMed]
17. Mohd Zaid, H.F.; Chong, F.K.; Abdul Mutalib, M.I. Photooxidative-extractive deep desulfurization of diesel using Cu–Fe/TiO_2 and eutectic ionic liquid. *Fuel* **2015**, *156*, 54–62. [CrossRef]
18. Lü, X.; Yang, W.; Quan, Z.; Lin, T.; Bai, L.; Wang, L.; Huang, F.; Zhao, Y. Enhanced Electron Transport in Nb-Doped TiO_2 Nanoparticles via Pressure-Induced Phase Transitions. *J. Am. Chem. Soc.* **2014**, *136*, 419–426. [CrossRef] [PubMed]
19. Williams, G.; Seger, B.; Kamat, P. TiO_2-Graphene Nanocomposites. UV-Assisted Photocatalytic Reduction of Graphene Oxide. *ACS Nano* **2008**, *2*, 1487–1491.
20. Yeh, T.F.; Cihlář, J.; Chang, C.Y.; Cheng, C.; Teng, H. Roles of graphene oxide in photocatalytic water splitting. *Mater. Today* **2013**, *16*, 78–84. [CrossRef]
21. White, R.; Thomas, P.S.; Philips, M.R.; Wuhrer, R.; Guerbois, J.P. TG-MS characterization of the reaction products of cadmium yellow and malachite artist's pigments. *J. Therm. Anal. Calorim.* **2007**, *88*, 181–184. [CrossRef]
22. Wang, J.; Qu, S.; Zhong, Z.; Wang, S.; Liu, K.; Hu, A. Fabrication of TiO_2 nanoparticles/nanorod composite arrays via a two-step method for efficient dye-sensitized solar cells. *Prog. Nat. Sci. Mater. Int.* **2014**, *24*, 588–592. [CrossRef]
23. Marcano, D.C.; Kosynkin, D.V.; Berlin, J.M.; Sinitskii, A.; Sun, Z.; Slesarev, A.; Alemany, L.B.; Lu, W.; Tour, J.M. Improved Synthesis of Graphene Oxide. *ACS Nano* **2010**, *4*. [CrossRef] [PubMed]
24. Frank, O.; Zukalova, M.; Laskova, B.; Kürti, J.; Koltai, J.; Kavan, L. Raman spectra of titanium dioxide (anatase, rutile) with identified oxygen isotopes (16, 17, 18). *Phys. Chem. Chem. Phys.* **2012**, *14*, 14567. [CrossRef] [PubMed]
25. De Waal, D.; Hutter, C. Vibrational spectra of two phases of copper pyrovanadate and some solid solutions of copper and magnesium pyrovanadate. *Mater. Res. Bull.* **1994**, *29*, 843–849. [CrossRef]
26. How, G.T.S.; Pandikumar, A.; Ming, H.N.; Ngee, L.H. Highly exposed {001} facets of titanium dioxide modified with reduced graphene oxide for dopamine sensing. *Sci. Rep.* **2014**, *4*, 2–9. [CrossRef] [PubMed]
27. Mishra, S.K.; Tripathi, S.N.; Choudhary, V.; Gupta, B.D. SPR based fibre optic ammonia gas sensor utilizing nanocomposite film of PMMA/reduced graphene oxide prepared by in situ polymerization. *Sens. Actuators B Chem.* **2014**, *199*, 190–200. [CrossRef]
28. Viezbicke, B.D.; Patel, S.; Davis, B.E.; Birnie, D.P. Evaluation of the Tauc method for optical absorption edge determination: ZnO thin films as a model system. *Phys. Status Solidi* **2015**, *252*, 1700–1710. [CrossRef]
29. Shuang, S.; Lv, R.; Xie, Z.; Wang, W.; Cui, X.; Ning, S.; Zhang, Z. α-Fe_2O_3 nanopillar arrays fabricated by electron beam evaporation for the photoassisted degradation of dyes with H_2O_2. *RSC Adv.* **2016**, *6*, 534–540. [CrossRef]
30. Li, H.; Su, Z.; Hu, S.; Yan, Y. Free-standing and flexible Cu/Cu_2O/CuO heterojunction net: A novel material as cost-effective and easily recycled visible-light photocatalyst. *Appl. Catal. B Environ.* **2017**, *207*, 134–142. [CrossRef]
31. Hernández, S.; Hidalgo, D.; Sacco, A.; Chiodoni, A.; Lamberti, A.; Cauda, V.; Tresso, E.; Saracco, G. Comparison of photocatalytic and transport properties of TiO_2 and ZnO nanostructures for solar-driven water splitting. *Phys. Chem. Chem. Phys.* **2015**, *17*, 7775–7786. [CrossRef] [PubMed]
32. Kwolek, P.; Szaciłowski, K. Photoelectrochemistry of n-type bismuth oxyiodide. *Electrochim. Acta* **2013**, *104*, 448–453. [CrossRef]
33. Pastor-Moreno, G. Chapter 5: Electrochemical studies of moderately boron doped diamond in non aqueous electrolyte. *Electrochem. Appl. CVD Diam.* **2002**, 2–4.
34. Yuan, W.; Yuan, J.; Xie, J.; Li, C.M. Polymer-Mediated Self-Assembly of TiO_2@Cu_2O Core-Shell Nanowire Array for Highly Efficient Photoelectrochemical Water Oxidation. *ACS Appl. Mater. Interfaces* **2016**, *8*, 6082–6092. [CrossRef] [PubMed]

35. Shuang, S.; Lv, R.; Xie, Z.; Zhang, Z. Surface Plasmon Enhanced Photocatalysis of Au/Pt-decorated TiO_2 Nanopillar Arrays. *Sci. Rep.* **2016**, *6*, 26670. [CrossRef] [PubMed]
36. Tu, N.D.K.; Choi, J.; Park, C.R.; Kim, H. Remarkable Conversion between n- and p-Type Reduced Graphene Oxide on Varying the Thermal Annealing Temperature. *Chem. Mater.* **2015**, *27*, 7362–7369. [CrossRef]
37. Moniz, S.J.A.; Tang, J. Charge transfer and photocatalytic activity in CuO/TiO_2 nanoparticle heterojunctions synthesised through a rapid, one-pot, microwave solvothermal route. *ChemCatChem* **2015**, *7*, 1659–1667. [CrossRef]
38. Singh, P.; Nath, P.; Arun, R.K.; Mandal, S.; Chanda, N. Novel synthesis of a mixed Cu/CuO-reduced graphene oxide nanocomposite with enhanced peroxidase-like catalytic activity for easy detection of glutathione in solution and using a paper strip. *RSC Adv.* **2016**, *6*, 92729–92738. [CrossRef]
39. Chen, W.; Yan, L.; Bangal, P.R. Preparation of graphene by the rapid and mild thermal reduction of graphene oxide induced by microwaves. *Carbon N. Y.* **2010**, *48*, 1146–1152. [CrossRef]
40. Pei, S.; Cheng, H.M. The reduction of graphene oxide. *Carbon N. Y.* **2012**, *50*, 3210–3228. [CrossRef]

Article

High Sensitivity Surface Plasmon Resonance Sensor Based on Two-Dimensional MXene and Transition Metal Dichalcogenide: A Theoretical Study

Yi Xu [1,2], Yee Sin Ang [1], Lin Wu [2] and Lay Kee Ang [1,*]

1 SUTD-MIT International Design Center & Science and Math Cluster, Singapore University of Technology and Design (SUTD), 8 Somapah Road, Singapore 487372, Singapore; yi_xu@mymail.sutd.edu.sg (Y.X.); yeesin_ang@sutd.edu.sg (Y.S.A.)
2 Institute of High Performance Computing, Agency for Science, Technology, and Research (A*STAR), 1 Fusionopolis Way, #16-16 Connexis, Singapore 138632, Singapore; wul@ihpc.a-star.edu.sg
* Correspondence: ricky_ang@sutd.edu.sg; Tel.: +65-6499-4558

Received: 23 November 2018; Accepted: 26 January 2019; Published: 29 January 2019

Abstract: MXene, a new class of two-dimensional nanomaterials, have drawn increasing attention as emerging materials for sensing applications. However, MXene-based surface plasmon resonance sensors remain largely unexplored. In this work, we theoretically show that the sensitivity of the surface plasmon resonance sensor can be significantly enhanced by combining two-dimensional $Ti_3C_2T_x$ MXene and transition metal dichalcogenides. A high sensitivity of 198°/RIU (refractive index unit) with a sensitivity enhancement of 41.43% was achieved in aqueous solutions (refractive index ~1.33) with the employment of monolayer $Ti_3C_2T_x$ MXene and five layers of WS_2 at a 633 nm excitation wavelength. The integration of $Ti_3C_2T_x$ MXene with a conventional surface plasmon resonance sensor provides a promising approach for bio- and chemical sensing, thus opening up new opportunities for highly sensitive surface plasmon resonance sensors using two-dimensional nanomaterials.

Keywords: MXene; $Ti_3C_2T_x$; transition metal dichalcogenides; surface plasmon resonance; sensitivity

1. Introduction

Optical sensors based on surface plasmon resonance (SPR) has been widely used for biosensing and chemical sensing in the past few decades due to their superior characteristics, such as being highly sensitive, reliable, label-free, and their capacity for real-time detection [1–5]. Various types of SPR sensors [1,2], including prism-coupled SPR sensors, metallic-grating coupled SPR sensors, fiber optic SPR sensors, and waveguide-based SPR sensors, have been designed and demonstrated for sensing applications. The Kretschmann configuration [6] is a typical prism-coupled SPR sensor structure, in which plasmonic metal (e.g., gold) film is deposited onto the base of a prism. A transverse magnetic (TM)-polarized incident light undergoes total internal reflection at the prism/metal film interface and generates an evanescent wave that penetrates through the metal thin film. Thus exciting a surface plasmon at the interface between the metal film and sensing medium (i.e., the outer boundary of metal film). The excitation of the surface plasmons results in a resonant dip in the angular spectrum of the reflected light with a fixed excitation light wavelength. The excitation of the surface plasmon depends on the refractive index (RI) of the sensing medium (or analyte), and a slight change in the analyte RI will produce a variation in the position (i.e., resonance angle) and magnitude of the resonance dip. This variation of resonance angle can be employed for the sensitive detection of RI change [1,2].

To obtain a highly sensitive SPR sensor, various techniques have been proposed and demonstrated [7], such as coating a dielectric material on the metal film [8]. In recent years, graphene, a two-dimensional (2D) nanomaterial, has been proposed and implemented to improve the sensitivities of SPR sensors [9–12] due to its unusual optical properties [13–19]. For example, Wu et al. [9] first proposed a graphene-based SPR biosensor consising of a graphene-on-Au structure. This graphene-integrated SPR sensor exhibited enhanced sensitivity, compared to the bare Au-based conventional SPR sensor, and a sensitivity enhancement of 25% was achieved with 10 layers of graphene applied. Besides graphene, SPR sensors with 2D transition metal dichalcogenides (TMDs), including molybdenum disulfide (MoS_2), molybdenum diselenide ($MoSe_2$), tungsten disulfide (WS_2), and tungsten diselenide (WSe_2), have been studied [20–25]. Ouyang et al. [20] theoretically investigated the sensor performances of TMDs-based SPR sensors with the structure of Au/Si/TMDs under different excitation wavelengths. The highest RI sensitivity of 155.68°/RIU (RIU: refractive index unit) was obtained with the 35 nm Au/7 nm Si/monolayer WS_2 structure at the wavelength of 600 nm. Another study on MoS_2-integrated SPR sensors has demonstrated that the MoS_2-based SPR sensor possesses better sensor performance (higher sensitivity and detection accuracy) than that of graphene-based sensors in the near-infrared regime [21].

MXenes [26–28], a new class of 2D materials consisting of transition metal carbides, nitrides, and carbonitrides, have attracted increasing attention in recent years due to their exceptional properties, including novel electrochemical properties [29] and extremely high electrical conductivity [30]. Furthermore, MXenes exhibit higly accessible hydrophilic surfaces [31], which is in contrast to graphene and most other 2D materials. Owing to their unique properties, MXenes have demonstrated promise for various applications, such as energy storage [31], water purification [32], chemical catalysts [33], photocatalysts [34], electrocatalysts [35], and photothermal therapy [36]. The MXene is also a promising material for sensing applications [37,38], such as electrochemical sensors [39,40], field effect transistor sensors [41], electrochemiluminescent sensors [42] and gas sensors [43,44]. For example, Kim et al. [44] recently demonstrated a $Ti_3C_2T_x$ MXene gas sensor by making use of its high metallic conductivity and fully functionalized surface. This $Ti_3C_2T_x$ MXene sensor exhibited higher sensitivity than that of gas sensors based on conventional semiconducting channel materials. It also possessed an ultra-high signal-to-noise ratio, which was two orders of magnitude greater than those of MoS_2, black phosphorus, and reduced graphene oxide integrated sensors. Lorencova et al. [45] proposed and demonstrated a $Ti_3C_2T_x$-based electrochemical sensor for H_2O_2 sensing. A detection limit of 0.7 nM was achieved, which is comparable to the best recorded so far (0.3 nM) [46]. However, few reports on MXene-integrated SPR sensors are available [47]. For example, a recent theoretical investigation on an $Ti_3C_2T_x$ MXene-based SPR sensor [47] showed that coating $Ti_3C_2T_x$ layers on Au film could enhance the sensitivity of a conventional Au-based SPR sensor. A RI sensitivity of 160°/RIU was achieved with four layers of $Ti_3C_2T_x$-coated Au film at a 633 nm excitation wavelength, whereas it was 137°/RIU for the $Ti_3C_2T_x$-devoid setup.

In this work, we designed a new MXene-based SPR sensor with the combination of $Ti_3C_2T_x$ MXene and TMDs. The resulting structure exhibited significantly improved sensitivity compared to the 2D materials-devoid setup. A highest RI sensitivity of 198°/RIU was achieved for the Au/five-layer-WS_2/Au/monolayer $Ti_3C_2T_x$ MXene structure in aqueous solutions with an excitation wavelength of 633 nm, which was a 41.43% sensitivity enhancement when compared with the conventional bare Au-based SPR sensor. The proposed MXene-TMDs plasmonic platform could offer new opportunities for highly sensitive SPR sensing. In addition, since the traditional prism-based SPR sensors have been successfully commericalized, such as Biacore (GE Healthcare), the proposed 2D nanomaterials-integrated SPR sensor could also stimulate new interest toward the exploration of commercially available high sensitivity SPR sensors.

2. Theoretical Model

The proposed SPR sensor structure is based on a modified Kretschmann configuration, as shown in Figure 1. In the proposed sensor structure, an Au film with the thickness of $d_2 = 50$ nm is attached to the base of a BK7 prism. Another thinner Au film ($d_4 = 10$ nm), decorated with TMDs and $Ti_3C_2T_x$

MXene on each side, is deposited on the previous thick Au film (see Figure 1). The $Ti_3C_2T_x$ MXene is kept in contact with the sensing medium or analyte. A TM-polarized light from a monochromatic source ($\lambda = 633$ nm) is launched in one side of the BK7 prism and the reflected light is detected from the other side. By scanning the incident angle to obtain an angular spectrum of the reflected light, and monitoring the resonance angle shift, the analyte RI variations can be observed.

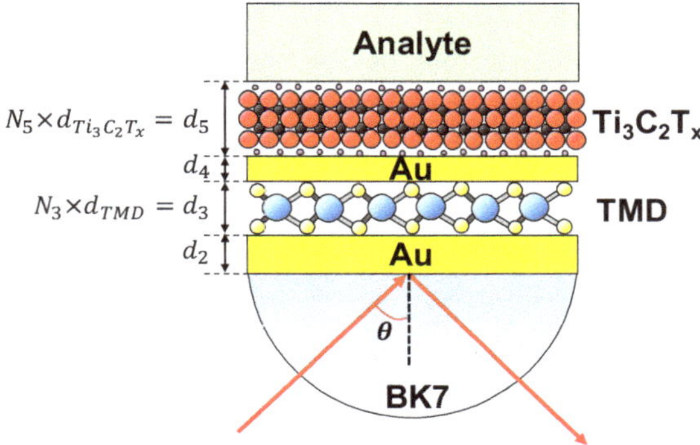

Figure 1. Schematic illustration of the proposed surface plasmon resonance (SPR) sensor with $Ti_3C_2T_x$ and 2D transition metal dichalcogenides (TMD) layers.

The reflectance R of the proposed sensor can be calculated with a generalized N-layer model [48]. The reflectance for the TM-polarized incident light is:

$$R = \left| \frac{(M_{11} + M_{12}q_N)q_1 - (M_{21} + M_{22}q_N)}{(M_{11} + M_{12}q_N)q_1 + (M_{21} + M_{22}q_N)} \right|^2, \tag{1}$$

in which M_{11}, M_{12}, M_{21}, and M_{22} are the four elements of the matrix M given by:

$$M = \begin{bmatrix} M_{11} & M_{12} \\ M_{21} & M_{22} \end{bmatrix} = \prod_{k=2}^{N-1} M_k, \tag{2}$$

with:

$$M_k = \begin{bmatrix} \cos\beta_k & -i(\sin\beta_k)/q_k \\ -iq_k \sin\beta_k & \cos\beta_k \end{bmatrix}. \tag{3}$$

Here,

$$\beta_k = \frac{2\pi d_k}{\lambda}\left(n_k^2 - n_1^2 \sin^2\theta_1\right)^{1/2}, \tag{4}$$

and

$$q_k = \frac{\left(n_k^2 - n_1^2 \sin^2\theta_1\right)^{1/2}}{n_k^2}, \tag{5}$$

in which λ is the wavelength of incident TM-polarized light, and θ_1 is the incident angle. d_k and n_k are the thickness and RI of the kth layer with $k = 2$ to $N-1$, respectively. The first layer ($k = 1$) in the sensor structure is the BK7 prism, and the wavelength-dependent RI is given by [49]:

$$n_{\text{BK7}} = \sqrt{1 + \frac{1.03961212\lambda^2}{\lambda^2 - 0.00600069867} + \frac{0.231792344\lambda^2}{\lambda^2 - 0.0200179144} + \frac{1.01046945\lambda^2}{\lambda^2 - 103.560653}}, \tag{6}$$

in which the wavelength λ is given in μm. The Nth layer is the analyte, and its RI is defined as $n_a = 1.33$ (water). The complex RI of Au film is calculated according to the Drude–Lorentz model [50]:

$$n_{\mathrm{Au}} = \sqrt{1 - \frac{\lambda^2 \lambda_c}{\lambda_p^2(\lambda_c + i\lambda)}}, \tag{7}$$

where λ_c (=8.9342 $\times 10^{-6}$ m) and λ_p (=1.6826 $\times 10^{-7}$ m) is the collision wavelength and the plasma wavelength of Au, respectively. Monolayer $Ti_3C_2T_x$ has a thickness of $d_{Ti_3C_2T_x} = 0.993$ nm [51], and its refractive index is $2.38 + 1.33i$ at the wavelength of 633 nm [52]. For monolayer TMD, the thickness is 0.65 nm, 0.7 nm, 0.8 nm and 0.7 nm for MoS_2, $MoSe_2$, WS_2 and WSe_2, respectively. And the corresponding complex RI at the wavelength of 633 nm is $5.0805 + 1.1723i$, $4.6226 + 1.0063i$, $4.8937 + 0.3124i$, and $4.5501 + 0.4332i$, respectively [23,53,54]. In the proposed sensor structure, the layer number of the TMD is N_3, and it is N_5 for $Ti_3C_2T_x$. The reflectance R depends on the analyte RI n_a, and a variation of analyte RI Δn_a will result in a change in the reflectance, as well as the resonance angle $\Delta\theta_{res}$. Therefore, the sensitivity is defined as:

$$S = \frac{\Delta\theta_{res}}{\Delta n_a}. \tag{8}$$

3. Results and Discussion

2D material-on-Au has been experimentally obtained in recent years. For example, graphene on Au surface has been experimentally demonstrated using the transfer printing technique [55,56]. The obtained graphene-on-Au structure was experimentally demonstrated for SPR sensing applications [56]. TMDs on the Au surface were also experimentally achieved [57–62]. These techniques can be applied for the fabrication of MXene-on-Au structures. Therefore, the proposed SPR sensor based on 2D MXene and TMDs are expected to be achieved easily. In order to illustrate the sensitivity enhancement of the proposed SPR sensor, we calculated the angular spectrum of the reflected light for various sensor structures, as shown in Figure 2, before (solid lines) and after (dashed lines) the RI variation of the sensing medium, assuming a small RI change $\Delta n_a = 0.005$. For each SPR sensor, the increase of the analyte RI will shift the resonance angle toward a larger value. For example, for the SPR sensor with $N_3 = 0$ and $N_5 = 0$ (i.e., conventional SPR sensor with 60 nm (=$d_2 + d_4$) Au film shown in Figure 2a), the resonance angle is 70.64° with the ambient RI of 1.330, and increases to 71.34° with a small analyte RI increment ($\Delta n_a = 0.005$). Therefore, a sensitivity of $S_0 = 140°$/RIU was obtained for the bare Au-based SPR sensor. By inserting a monolayer MoS_2 between the two Au films (i.e., $N_3 = 1$ and $N_5 = 0$), an enhanced sensitivity of $S = 146°$/RIU was achieved (see Figure 2b). To study the sensitivity improvement with reference to the sensitivity of the conventional Au-based SPR sensor, we denoted the sensitivity enhancement as $(S - S_0)/S_0 \times 100\%$, in which S is the sensitivity of 2D-nanomaterial-integrated SPR sensor. For the SPR sensor shown in Figure 2b, a relatively low sensitivity enhancement of 4.29% was obtained. The sensitivity and sensitivity enhancement were improved to 150°/RIU and 7.14%, respectively, with only one layer of $Ti_3C_2T_x$ (i.e., $N_3 = 0$, $N_5 = 1$, Figure 2c). With the employment of both a $Ti_3C_2T_x$ MXene and MoS_2 layer ($N_3 = 1$ and $N_5 = 1$), an enhanced sensitivity of $S = 156°$/RIU with the sensitivity enhancement of 11.43% was achieved, as shown in Figure 2d. Besides the $Ti_3C_2T_x$-MoS_2-based SPR sensor, three other TMDs ($MoSe_2$, WS_2, WSe_2) and $Ti_3C_2T_x$ integrated SPR sensors ($N_3 = 1$ and $N_5 = 1$) also exhibited enhanced sensitivity (Figures S1–S3 in the Supporting Information). Therefore, the proposed SPR sensor with the simultaneous employment of $Ti_3C_2T_x$ and TMDs exhibited enhanced sensitivity and offers the potential for highly sensitive sensing applications.

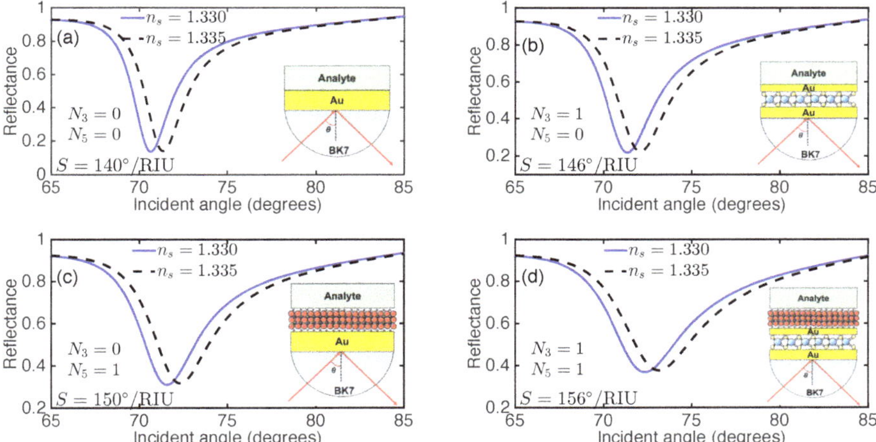

Figure 2. Reflectance as a function of the incident angle before (solid lines) and after (dashed lines) the variation of analyte refractive index (RI) for the $Ti_3C_2T_x$-MoS_2-based SPR sensor with (**a**) $N_3 = 0$, $N_5 = 0$, (i.e., no 2D materials); (**b**) $N_3 = 1$, $N_5 = 0$, (i.e., monolayer MoS_2); (**c**) $N_3 = 0$, $N_5 = 1$, (i.e., monolayer $Ti_3C_2T_x$), and (**d**) $N_3 = 1$, $N_5 = 1$, (i.e., monolayer MoS_2 and monolayer $Ti_3C_2T_x$).

The study above only focuses on monolayer MoS_2 and $Ti_3C_2T_x$. Previous investigations on 2D-material-integrated SPR sensors have demonstrated that the sensitivity also depends on the layer number of 2D materials [9–11,20–24]. Therefore, it is necessary to study the effect of number of $Ti_3C_2T_x$ and MoS_2 layers on the sensitivity. First, we investigated the effect of multiple layers of 2D materials on the reflectance for the proposed SPR sensor. The reflectance as a function of the incident angle for the monolayer $Ti_3C_2T_x$-MoS_2-based SPR sensor with different numbers of MoS_2 layers is shown in Figure 3a. It was readily apparent that the resonance angle increased with the number of MoS_2 layers due to the increased propagation constant (wavevector) of the surface plasmons. In addition, a shallowing and broadening of the reflectance curves was observed when the layers of MoS_2 increased, due to the increased electron energy loss [20,22]. Similar phenomena were found in the reflectance curves for $Ti_3C_2T_x$-monolayer MoS_2-based SPR sensors with different numbers of $Ti_3C_2T_x$ layers, as shown in Figure 3b. By comparing Figure 3a and Figure 3b, it was found that the increased energy loss caused by the integration of $Ti_3C_2T_x$ layers was larger than that caused by the additional MoS_2 layers.

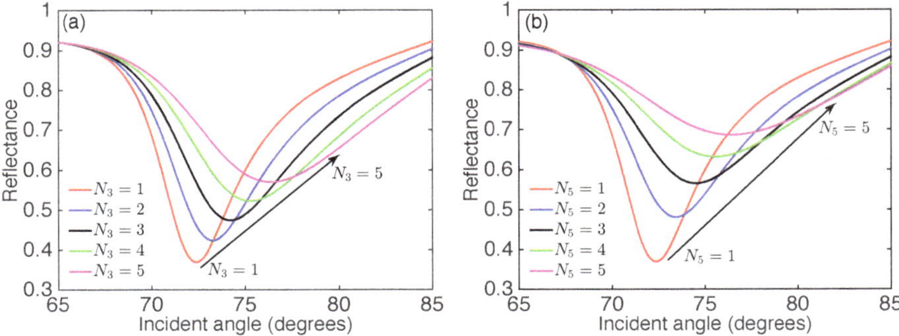

Figure 3. Reflectance as a function of the incident angle for $Ti_3C_2T_x$-MoS_2-based SPR sensor with (**a**) different number of MoS_2 (N_3) and monolayer $Ti_3C_2T_x$ ($N_5 = 1$), and (**b**) different number of $Ti_3C_2T_x$ (N_5) and monolayer MoS_2 ($N_3 = 1$).

To further improve the sensitivity of proposed SPR sensor, we studied the optimiziation of the sensitivity by varying the layer number of the $Ti_3C_2T_x$ MXene and TMDs. The sensitivity as a function of the number of MoS_2 layers for the $Ti_3C_2T_x$-MoS_2-based SPR sensor with different numbers of $Ti_3C_2T_x$ layers is shown in Figure 4. The sensitivity first increased and then decreased with the number of MoS_2 layers, when the SPR sensor integrated monolayer and two layers of $Ti_3C_2T_x$. However, adding more layers of $Ti_3C_2T_x$ (e.g., three to five layers) resulted in decreased sensitivity with the number of MoS_2 layers. Due to the relative higher energy loss of the $Ti_3C_2T_x$ layers, the SPR signal enhancement effect of the MoS_2 layers in the SPR sensor with three to five layers of $Ti_3C_2T_x$ was overwhelmed by the energy loss with the additional MoS_2 layers. In contrast, with the integration of monolayer $Ti_3C_2T_x$, the sensitivity increased with the number of MoS_2 layers from one to four (see Figure 4), where the SPR signal enhancement effect was more significant than the energy loss caused by the MoS_2 layers [22]. The maximum sensitivity of 174°/RIU was found for the $Ti_3C_2T_x$-MoS_2-based SPR sensor integrated with four-layer MoS_2 and monolayer $Ti_3C_2T_x$.

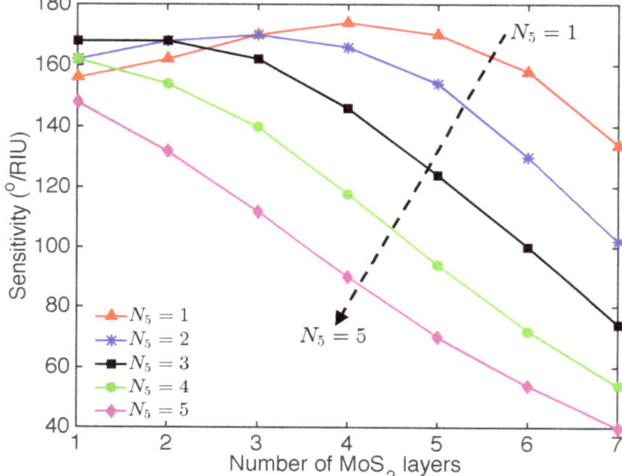

Figure 4. Sensitivity as a function of the number of MoS_2 layers for $Ti_3C_2T_x$-MoS_2-based SPR sensor with different layers of $Ti_3C_2T_x$.

The optimization of various combinations of $Ti_3C_2T_x$ MXene and TMDs (e.g., $Ti_3C_2T_x$-$MoSe_2$, $Ti_3C_2T_x$-WS_2, and $Ti_3C_2T_x$-WSe_2) of the SPR sensors are shown in Figures S4–S6 of the Supporting Information. It was found that only monolayer $Ti_3C_2T_x$ MXene could be used to obtain the maximum sensitivity for the $Ti_3C_2T_x$-TMDs-based SPR sensors. The sensitivity and sensitivity enhancement at the optimized number of TMD layers and $Ti_3C_2T_x$ MXene layers for the proposed SPR sensor structure are summarized in Table 1. The $Ti_3C_2T_x$-WS_2- and $Ti_3C_2T_x$-WSe_2-based SPR sensors possessed sensitivities more than 190°/RIU. A maximum sensitivity of 198°/RIU was achieved with the sensor structure of Au/WSe_2 (six layers)/$Ti_3C_2T_x$ (one layer)/Au, and a sensitivity enhancement of 41.43% was obtained. The sensitivities achieved with the proposed $Ti_3C_2T_x$-TMDs-based SPR sensors at a 633 nm excitation wavelength were significantly higher than that of the conventional Au-$Ti_3C_2T_x$ (four layer)-based SPR sensor (160°/RIU) recently reported by Wu et al. [47]. The combination of TMDs and $Ti_3C_2T_x$ offers the alternative of sensitivity enhancement for $Ti_3C_2T_x$-based SPR sensors.

Table 1. Sensitivity and sensitivity enhancement at the optimized number of TMD layers and $Ti_3C_2T_x$ layers for the $Ti_3C_2T_x$-TMDs-based SPR sensor.

Type of TMD	Number of TMD Layers N_3	Number of $Ti_3C_2T_x$ Layers N_5	Sensitivity (°/RIU)	$(S - S_0)/S_0$ (%)
MoS_2	4	1	174	24.29
$MoSe_2$	5	1	176	25.71
WS_2	5	1	198	41.43
WSe_2	6	1	192	37.14

The RI of the surrounding environment was also important to the sensitivity, which determined the appropriate working RI range or working environment (e.g., gas or liquid) of the SPR sensor. The sensitivity for the optimized $Ti_3C_2T_x$-TMDs-based SPR sensor was plotted with varying analyte RI in Figure 5. The optimized $Ti_3C_2T_x$-TMDs-based SPR sensor possessed a relatively low sensitivity (<90°/RIU) within the analyte RI range from 1.0 to 1.15. This revealed that the proposed SPR sensor was not appropriate for gas sensing, which typically involves a RI $\sim$ 1.0. The sensitivity of the optimized SPR sensor first increased to a maximum and then decreased with the analyte RI in the range of 1.0–1.36. The maximum RI sensitivity was found around the analyte RI of 1.330 (i.e., the RI of water). Therefore, the proposed sensor was more suited for operating in an aqueous medium, particularly for bio- and chemical sensing.

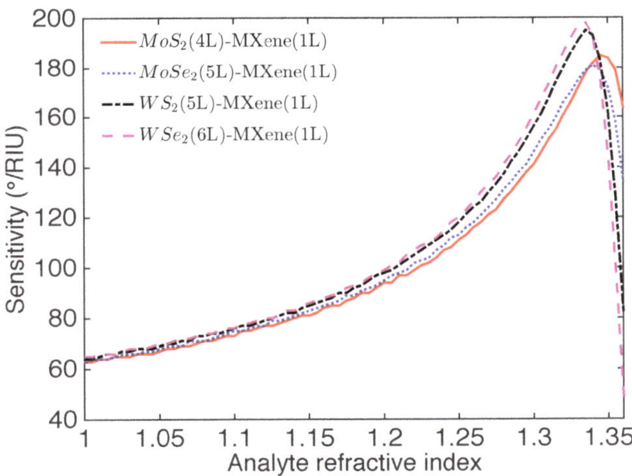

Figure 5. Variation of sensitivity for the optimized $Ti_3C_2T_x$-TMD-based SPR sensor with the varying analyte RI.

4. Conclusions

A novel SPR sensor based on Au-$Ti_3C_2T_x$-Au-TMDs is theoretically presented. The MXene-TMDs-integrated SPR sensor possessed enhanced sensitivity as compared to the bare Au film-based SPR sensor. For the aqueous solutions (RI $\sim$1.33), the RI sensitivities of 174°/RIU, 176°/RIU, 198°/RIU, and 192°/RIU for the proposed SPR sensor with monolayer $Ti_3C_2T_x$ MXene and four-layer MoS_2, five-layer $MoSe_2$, five-layer WS_2, and six-layer WSe_2, respectively, were achieved at the 633 nm excitation wavelength. Compared to the conventional Au film SPR sensor, the sensitivity was significantly enhanced by 24.29%, 25.71%, 41.43%, and 37.14%, respectively. The high sensitivities of the proposed $Ti_3C_2T_x$ MXene-based SPR sensors offer a potential route towards highly sensitive SPR sensors. Although this work was purely based on theoretical calculations, we used realistic material parameters and the results could be readily verified by experimental investigations. Moreover, since the structures of graphene-on-Au and TMDs-on-Au have been experimentally realized in recent years [55–62], it is possible to fabricate the MXene-on-Au structure. Thus the proposed SPR sensor based on 2D MXene and TMDs is experimentally feasible.

Supplementary Materials: The following are available online at http://www.mdpi.com/2079-4991/9/2/165/s1, Figure S1: Reflectance for the $Ti_3C_2T_x$-$MoSe_2$-based SPR sensor; Figure S2: Reflectance for the $Ti_3C_2T_x$-WS_2-based SPR sensor; Figure S3: Reflectance for the $Ti_3C_2T_x$-WSe_2-based SPR sensor; Figure S4: Variation of sensitivity with number of $MoSe_2$ and $Ti_3C_2T_x$ layers for the $Ti_3C_2T_x$-$MoSe_2$-based SPR sensor; Figure S5: Variation of sensitivity with number of WS_2 and $Ti_3C_2T_x$ layers for the $Ti_3C_2T_x$-WS_2-based SPR sensor; Figure S6: Variation of sensitivity with number of WSe_2 and $Ti_3C_2T_x$ layers for the $Ti_3C_2T_x$-WSe_2-based SPR sensor.

Author Contributions: Conceptualization, Y.X.; investigation, Y.X., Y.S.A., L.W. and L.K.A; writing-original draft preparation, Y.X.; writing-review and editing, Y.X., Y.S.A., L.W. and L.K.A.; supervision, L.K.A.; funding acquisition, L.K.A.

Funding: This work was partially supported by the Singapore A*STAR AME IRG (A1783c0011) and USA Air Force Office of Scientific Research (AFOSR) through the Asian Office of Aerospace Research and Development (AOARD) under Grant No. FA2386-17-1-4020.

Conflicts of Interest: The authors declare no conflict of interest.

References

1. Homola, J.; Yee, S.S.; Gauglitz, G. Surface plasmon resonance sensors: Review. *Sens. Actuators B Chem.* **1999**, *54*, 3–15. [CrossRef]
2. Homola, J. Surface plasmon resonance sensors for detection of chemical and biological species. *Chem. Rev.* **2008**, *108*, 462–493. [CrossRef] [PubMed]
3. Fan, X.; White, I.M.; Shopova, S.I.; Zhu, H.; Suter, J.D.; Sun, Y. Sensitive optical biosensors for unlabeled targets: A review. *Anal. Chim. Acta* **2008**, *620*, 8–26. [CrossRef] [PubMed]
4. Wijaya, E.; Lenaerts, C.; Maricot, S.; Hastanin, J.; Habraken, S.; Vilcot, J.P.; Boukherroub, R.; Szunerits, S. Surface plasmon resonance-based biosensors: From the development of different SPR structures to novel surface functionalization strategies. *Curr. Opin. Solid State Mater. Sci.* **2011**, *15*, 208–224. [CrossRef]
5. Masson, J.F. Surface plasmon resonance clinical biosensors for medical diagnostics. *ACS Sens.* **2017**, *2*, 16–30. [CrossRef] [PubMed]
6. Kretschmann, E.; Raether, H. Notizen: Radiative decay of non radiative surface plasmons excited by light. *Z. Naturforschung A* **1968**, *23*, 2135–2136. [CrossRef]
7. Shalabney, A.; Abdulhalim, I. Sensitivity-enhancement methods for surface plasmon sensors. *Laser Photonics Rev.* **2011**, *5*, 571–606. [CrossRef]
8. Bhatia, P.; Gupta, B.D. Surface-plasmon-resonance-based fiber-optic refractive index sensor: Sensitivity enhancement. *Appl. Opt.* **2011**, *50*, 2032–2036. [CrossRef]
9. Wu, L.; Chu, H.; Koh, W.; Li, E. Highly sensitive graphene biosensors based on surface plasmon resonance. *Opt. Express* **2010**, *18*, 14395–14400. [CrossRef] [PubMed]
10. Verma, R.; Gupta, B.D.; Jha, R. Sensitivity enhancement of a surface plasmon resonance based biomolecules sensor using graphene and silicon layers. *Sens. Actuators B Chem.* **2011**, *160*, 623–631. [CrossRef]
11. Fu, H.; Zhang, S.; Chen, H.; Weng, J. Graphene enhances the sensitivity of fiber-optic surface plasmon resonance biosensor. *IEEE Sens. J.* **2015**, *15*, 5478–5482. [CrossRef]
12. Wei, W.; Nong, J.; Zhu, Y.; Zhang, G.; Wang, N.; Luo, S.; Chen, N.; Lan, G.; Chuang, C.J.; Huang, Y. Graphene/Au-enhanced plastic clad silica fiber optic surface plasmon resonance sensor. *Plasmonics* **2018**, *13*, 483–491. [CrossRef]
13. Ju, L.; Geng, B.; Horng, J.; Girit, C.; Martin, M.; Hao, Z.; Bechtel, H.A.; Liang, X.; Zettl, A.; Shen, Y.R.; et al. Graphene plasmonics for tunable terahertz metamaterials. *Nat. Nanotechnol.* **2011**, *6*, 630–634. [CrossRef] [PubMed]
14. Rodrigo, D.; Limaj, O.; Janner, D.; Etezadi, D.; De Abajo, F.J.G.; Pruneri, V.; Altug, H. Mid-infrared plasmonic biosensing with graphene. *Science* **2015**, *349*, 165–168. [CrossRef] [PubMed]
15. Huang, S.; Song, C.; Zhang, G.; Yan, H. Graphene plasmonics: Physics and potential applications. *Nanophotonics* **2016**, *6*, 1191–1204. [CrossRef]

16. Ang, Y.S.; Sultan, S.; Zhang, C. Nonlinear optical spectrum of bilayer graphene in the terahertz regime. *Appl. Phys. Lett.* **2010**, *97*, 243110. [CrossRef]
17. Ang, Y.S.; Chen, Q.; Zhang, C. Nonlinear optical response of graphene in terahertz and near-infrared frequency regime. *Front. Optoelectron.* **2015**, *8*, 3–26. [CrossRef]
18. Ooi, K.J.; Tan, D.T. Nonlinear graphene plasmonics. *Proc. R. Soc. A: Math. Phys. Eng. Sci.* **2017**, *473*, 20170433. [CrossRef]
19. Ooi, K.J.; Ang, Y.S.; Cheng, J.L.; Ang, L.K.; Tan, D.T. Electronic scattering of graphene plasmons in the terahertz nonlinear regime. *IEEE J. Sel. Top. Quantum Electron.* **2017**, *23*, 1–6. [CrossRef]
20. Ouyang, Q.; Zeng, S.; Jiang, L.; Hong, L.; Xu, G.; Dinh, X.Q.; Qian, J.; He, S.; Qu, J.; Coquet, P.; et al. Sensitivity enhancement of transition metal dichalcogenides/silicon nanostructure-based surface plasmon resonance biosensor. *Sci. Rep.* **2016**, *6*, 28190. [CrossRef]
21. Xu, Y.; Wu, L.; Ang, L.K. MoS_2-based Highly Sensitive Near-infrared Surface Plasmon Resonance Refractive index Sensor. *IEEE J. Sel. Top. Quantum Electron.* **2019**, *25*, 4600307. [CrossRef]
22. Zeng, S.; Hu, S.; Xia, J.; Anderson, T.; Dinh, X.Q.; Meng, X.M.; Coquet, P.; Yong, K.T. Graphene–MoS_2 hybrid nanostructures enhanced surface plasmon resonance biosensors. *Sens. Actuators B Chem.* **2015**, *207*, 801–810. [CrossRef]
23. Ouyang, Q.; Zeng, S.; Jiang, L.; Qu, J.; Dinh, X.Q.; Qian, J.; He, S.; Coquet, P.; Yong, K.T. Two-Dimensional Transition Metal Dichalcogenide Enhanced Phase-Sensitive Plasmonic Biosensors: Theoretical Insight. *J. Phys. Chem. C* **2017**, *121*, 6282–6289. [CrossRef]
24. Xu, Y.; Hsieh, C.Y.; Wu, L.; Ang, L.K. Two-dimensional transition metal dichalcogenides mediated long range surface plasmon resonance biosensors. *J. Phys. D Appl. Phys.* **2019**, *52*, 065101. [CrossRef]
25. Lin, Z.; Jiang, L.; Wu, L.; Guo, J.; Dai, X.; Xiang, Y.; Fan, D. Tuning and sensitivity enhancement of surface plasmon resonance biosensor with graphene covered Au-MoS_2-Au Films. *IEEE Photonics J.* **2016**, *8*, 1–8.
26. Naguib, M.; Kurtoglu, M.; Presser, V.; Lu, J.; Niu, J.; Heon, M.; Hultman, L.; Gogotsi, Y.; Barsoum, M.W. Two-dimensional nanocrystals produced by exfoliation of Ti3AlC2. *Adv. Mater.* **2011**, *23*, 4248–4253. [CrossRef] [PubMed]
27. Naguib, M.; Mashtalir, O.; Carle, J.; Presser, V.; Lu, J.; Hultman, L.; Gogotsi, Y.; Barsoum, M.W. Two-dimensional transition metal carbides. *ACS Nano* **2012**, *6*, 1322–1331. [CrossRef]
28. Naguib, M.; Mochalin, V.N.; Barsoum, M.W.; Gogotsi, Y. MXenes: A new family of two-dimensional materials. *Adv. Mater.* **2014**, *26*, 992–1005. [CrossRef] [PubMed]
29. Lukatskaya, M.R.; Kota, S.; Lin, Z.; Zhao, M.Q.; Shpigel, N.; Levi, M.D.; Halim, J.; Taberna, P.L.; Barsoum, M.W.; Simon, P.; et al. Ultra-high-rate pseudocapacitive energy storage in two-dimensional transition metal carbides. *Nat. Energy* **2017**, *2*, 17105. [CrossRef]
30. Lipatov, A.; Alhabeb, M.; Lukatskaya, M.R.; Boson, A.; Gogotsi, Y.; Sinitskii, A. Effect of synthesis on quality, electronic properties and environmental stability of individual monolayer Ti_3C_2 MXene flakes. *Adv. Electron. Mater.* **2016**, *2*, 1600255. [CrossRef]
31. Anasori, B.; Lukatskaya, M.R.; Gogotsi, Y. 2D metal carbides and nitrides (MXenes) for energy storage. *Nat. Rev. Mater.* **2017**, *2*, 16098. [CrossRef]
32. Peng, Q.; Guo, J.; Zhang, Q.; Xiang, J.; Liu, B.; Zhou, A.; Liu, R.; Tian, Y. Unique lead adsorption behavior of activated hydroxyl group in two-dimensional titanium carbide. *J. Am. Chem. Soc.* **2014**, *136*, 4113–4116. [CrossRef] [PubMed]
33. Xie, X.; Xue, Y.; Li, L.; Chen, S.; Nie, Y.; Ding, W.; Wei, Z. Surface Al leached Ti3AlC2 as a substitute for carbon for use as a catalyst support in a harsh corrosive electrochemical system. *Nanoscale* **2014**, *6*, 11035–11040. [CrossRef] [PubMed]
34. Mashtalir, O.; Cook, K.M.; Mochalin, V.; Crowe, M.; Barsoum, M.W.; Gogotsi, Y. Dye adsorption and decomposition on two-dimensional titanium carbide in aqueous media. *J. Mater. Chem. A* **2014**, *2*, 14334–14338. [CrossRef]
35. Seh, Z.W.; Fredrickson, K.D.; Anasori, B.; Kibsgaard, J.; Strickler, A.L.; Lukatskaya, M.R.; Gogotsi, Y.; Jaramillo, T.F.; Vojvodic, A. Two-dimensional molybdenum carbide (MXene) as an efficient electrocatalyst for hydrogen evolution. *ACS Energy Lett.* **2016**, *1*, 589–594. [CrossRef]

36. Xuan, J.; Wang, Z.; Chen, Y.; Liang, D.; Cheng, L.; Yang, X.; Liu, Z.; Ma, R.; Sasaki, T.; Geng, F. Organic-Base-Driven Intercalation and Delamination for the Production of Functionalized Titanium Carbide Nanosheets with Superior Photothermal Therapeutic Performance. *Angew. Chem.* **2016**, *128*, 14789–14794. [CrossRef]
37. Sinha, A.; Dhanjai; Zhao, H.; Huang, Y.; Lu, X.; Chen, J.; Jain, R. MXene: An emerging material for sensing and biosensing. *TrAC Trends Anal. Chem.* **2018**, *105*, 424–435. [CrossRef]
38. Zhu, J.; Ha, E.; Zhao, G.; Zhou, Y.; Huang, D.; Yue, G.; Hu, L.; Sun, N.; Wang, Y.; Lee, L.Y.S.; et al. Recent advance in MXenes: A promising 2D material for catalysis, sensor and chemical adsorption. *Coord. Chem. Rev.* **2017**, *352*, 306–327. [CrossRef]
39. Zhu, X.; Liu, B.; Hou, H.; Huang, Z.; Zeinu, K.M.; Huang, L.; Yuan, X.; Guo, D.; Hu, J.; Yang, J. Alkaline intercalation of Ti_3C_2 MXene for simultaneous electrochemical detection of Cd(II), Pb(II), Cu(II) and Hg(II). *Electrochim. Acta* **2017**, *248*, 46–57. [CrossRef]
40. Rasheed, P.A.; Pandey, R.P.; Rasool, K.; Mahmoud, K.A. Ultra-sensitive electrocatalytic detection of bromate in drinking water based on Nafion/$Ti_3C_2T_x$ (MXene) modified glassy carbon electrode. *Sens. Actuators B Chem.* **2018**, *265*, 652–659. [CrossRef]
41. Xu, B.; Zhu, M.; Zhang, W.; Zhen, X.; Pei, Z.; Xue, Q.; Zhi, C.; Shi, P. Ultrathin MXene-Micropattern-Based Field-Effect Transistor for Probing Neural Activity. *Adv. Mater.* **2016**, *28*, 3333–3339. [CrossRef]
42. Fang, Y.; Yang, X.; Chen, T.; Xu, G.; Liu, M.; Liu, J.; Xu, Y. Two-dimensional titanium carbide (MXene)-based solid-state electrochemiluminescent sensor for label-free single-nucleotide mismatch discrimination in human urine. *Sens. Actuators B Chem.* **2018**, *263*, 400–407. [CrossRef]
43. Lee, E.; VahidMohammadi, A.; Prorok, B.C.; Yoon, Y.S.; Beidaghi, M.; Kim, D.J. Room temperature gas sensing of two-dimensional titanium carbide (MXene). *ACS Appl. Mater. Interfaces* **2017**, *9*, 37184–37190. [CrossRef] [PubMed]
44. Kim, S.J.; Koh, H.J.; Ren, C.E.; Kwon, O.; Maleski, K.; Cho, S.Y.; Anasori, B.; Kim, C.K.; Choi, Y.K.; Kim, J.; et al. Metallic $Ti_3C_2T_x$ MXene Gas Sensors with Ultrahigh Signal-to-Noise Ratio. *ACS Nano* **2018**, *12*, 986–993. [CrossRef]
45. Lorencová, L.; Bertok, T.; Dosekova, E.; Holazová, A.; Paprckova, D.; Vikartovská, A.; Sasinková, V.; Filip, J.; Kasák, P.; Jerigová, M.; et al. Electrochemical performance of $Ti_3C_2T_x$ MXene in aqueous media: towards ultrasensitive H2O2 sensing. *Electrochim. Acta* **2017**, *235*, 471–479. [CrossRef]
46. Xiao, F.; Zhao, F.; Zhang, Y.; Guo, G.; Zeng, B. Ultrasonic electrodeposition of gold-platinum alloy nanoparticles on ionic liquid-chitosan composite film and their application in fabricating nonenzyme hydrogen peroxide sensors. *J. Phys. Chem. C* **2008**, *113*, 849–855. [CrossRef]
47. Wu, L.; You, Q.; Shan, Y.; Gan, S.; Zhao, Y.; Dai, X.; Xiang, Y. Few-layer $Ti_3C_2T_x$ MXene: A promising surface plasmon resonance biosensing material to enhance the sensitivity. *Sens. Actuators B Chem.* **2018**, *277*, 210–215. [CrossRef]
48. Yamamoto, M. Surface plasmon resonance (SPR) theory: Tutorial. *Rev. Polarogr.* **2002**, *48*, 209–237. [CrossRef]
49. Polyanskiy, M. Refractive Index Database. Available online: http://refractiveindex.info (accessed on 23 November 2018).
50. Maharana, P.K.; Srivastava, T.; Jha, R. On the performance of highly sensitive and accurate graphene-on-aluminum and silicon-based SPR biosensor for visible and near infrared. *Plasmonics* **2014**, *9*, 1113–1120. [CrossRef]
51. Shi, C.; Beidaghi, M.; Naguib, M.; Mashtalir, O.; Gogotsi, Y.; Billinge, S.J. Structure of nanocrystalline Ti_3C_2 MXene using atomic pair distribution function. *Phys. Rev. Lett.* **2014**, *112*, 125501. [CrossRef]
52. Miranda, A.; Halim, J.; Lorke, A.; Barsoum, M. Rendering $Ti_3C_2T_x$ (MXene) monolayers visible. *Mater. Res. Lett.* **2017**, *5*, 322–328. [CrossRef]
53. Li, Y.; Chernikov, A.; Zhang, X.; Rigosi, A.; Hill, H.M.; van der Zande, A.M.; Chenet, D.A.; Shih, E.M.; Hone, J.; Heinz, T.F. Measurement of the optical dielectric function of monolayer transition-metal dichalcogenides: MoS_2, $MoSe_2$, WS_2, and WSe_2. *Phys. Rev. B* **2014**, *90*, 205422. [CrossRef]
54. Liu, H.L.; Shen, C.C.; Su, S.H.; Hsu, C.L.; Li, M.Y.; Li, L.J. Optical properties of monolayer transition metal dichalcogenides probed by spectroscopic ellipsometry. *Appl. Phys. Lett.* **2014**, *105*, 201905. [CrossRef]
55. Song, B.; Li, D.; Qi, W.; Elstner, M.; Fan, C.; Fang, H. Graphene on Au(111): A highly conductive material with excellent adsorption properties for high-resolution bio/nanodetection and identification. *ChemPhysChem* **2010**, *11*, 585–589. [CrossRef] [PubMed]

56. Salihoglu, O.; Balci, S.; Kocabas, C. Plasmon-polaritons on graphene-metal surface and their use in biosensors. *Appl. Phys. Lett.* **2012**, *100*, 213110. [CrossRef]
57. Grønborg, S.S.; Ulstrup, S.; Bianchi, M.; Dendzik, M.; Sanders, C.E.; Lauritsen, J.V.; Hofmann, P.; Miwa, J.A. Synthesis of epitaxial single-layer MoS_2 on Au(111). *Langmuir* **2015**, *31*, 9700–9706. [CrossRef] [PubMed]
58. Sørensen, S.G.; Füchtbauer, H.G.; Tuxen, A.K.; Walton, A.S.; Lauritsen, J.V. Structure and electronic properties of in situ synthesized single-layer MoS_2 on a gold surface. *ACS Nano* **2014**, *8*, 6788–6796. [CrossRef] [PubMed]
59. Dendzik, M.; Michiardi, M.; Sanders, C.; Bianchi, M.; Miwa, J.A.; Grønborg, S.S.; Lauritsen, J.V.; Bruix, A.; Hammer, B.; Hofmann, P. Growth and electronic structure of epitaxial single-layer WS2 on Au(111). *Phys. Rev. B* **2015**, *92*, 245442. [CrossRef]
60. Bruix, A.; Miwa, J.A.; Hauptmann, N.; Wegner, D.; Ulstrup, S.; Grønborg, S.S.; Sanders, C.E.; Dendzik, M.; Čabo, A.G.; Bianchi, M.; et al. Single-layer MoS_2 on Au(111): Band gap renormalization and substrate interaction. *Phys. Rev. B* **2016**, *93*, 165422. [CrossRef]
61. Park, S.; Mutz, N.; Schultz, T.; Blumstengel, S.; Han, A.; Aljarb, A.; Li, L.J.; List-Kratochvil, E.J.; Amsalem, P.; Koch, N. Direct determination of monolayer MoS_2 and WSe_2 exciton binding energies on insulating and metallic substrates. *2D Mater.* **2018**, *5*, 025003. [CrossRef]
62. Lu, J.; Bao, D.L.; Qian, K.; Zhang, S.; Chen, H.; Lin, X.; Du, S.X.; Gao, H.J. Identifying and Visualizing the Edge Terminations of Single-Layer MoSe2 Island Epitaxially Grown on Au(111). *ACS Nano* **2017**, *11*, 1689–1695. [CrossRef] [PubMed]

Article

Facile and Controllable Synthesis of Large-Area Monolayer WS_2 Flakes Based on WO_3 Precursor Drop-Casted Substrates by Chemical Vapor Deposition

Biao Shi [1,†], Daming Zhou [1,†], Shaoxi Fang [1], Khouloud Djebbi [1,2], Shuanglong Feng [1], Hongquan Zhao [1], Chaker Tlili [1,*] and Deqiang Wang [1,2,*]

1 Chongqing Key Lab of Multi-Scale Manufacturing Technology, Chongqing Institute of Green and Intelligent Technology, Chinese Academy of Sciences, Chongqing 400714, China; shibiao@cigit.ac.cn (B.S.); dmzhou@cigit.ac.cn (D.Z.); fangshaoxi@cigit.ac.cn (S.F.); khouloud@cigit.ac.cn (K.D.); fengshuanglong@cigit.ac.cn (S.F.); hqzhao@cigit.ac.cn (H.Z.)

2 University of Chinese Academy of Sciences, Beijing 100049, China

* Correspondence: chakertlili@cigit.ac.cn (C.T.); dqwang@cigit.ac.cn (D.W.)

† These authors contributed equally to this work.

Received: 26 February 2019; Accepted: 2 April 2019; Published: 9 April 2019

Abstract: Monolayer WS_2 (Tungsten Disulfide) with a direct-energy gap and excellent photoluminescence quantum yield at room temperature shows potential applications in optoelectronics. However, controllable synthesis of large-area monolayer WS_2 is still challenging because of the difficulty in controlling the interrelated growth parameters. Herein, we report a facile and controllable method for synthesis of large-area monolayer WS_2 flakes by direct sulfurization of powdered WO_3 (Tungsten Trioxide) drop-casted on SiO_2/Si substrates in a one-end sealed quartz tube. The samples were thoroughly characterized by an optical microscope, atomic force microscope, transmission electron microscope, fluorescence microscope, photoluminescence spectrometer, and Raman spectrometer. The obtained results indicate that large triangular monolayer WS_2 flakes with an edge length up to 250 to 370 μm and homogeneous crystallinity were readily synthesized within 5 min of growth. We demonstrate that the as-grown monolayer WS_2 flakes show distinctly size-dependent fluorescence emission, which is mainly attributed to the heterogeneous release of intrinsic tensile strain after growth.

Keywords: WS_2; 2D materials; large-area; CVD; fluorescence emission; Raman mapping

1. Introduction

The isolation and synthesis of atomically thin two-dimensional transition metal dichalcogenides (TMDs), such as MS_2 (M = Mo, W), have attracted a lot of interest due to their unprecedented properties compared with their bulk counterparts [1–3]. Unlike graphene and h-BN (h-Born Nitride), atomically thin MS_2 possess semiconducting behavior with an intrinsic direct-energy gap corresponding to the visible frequency range, and show strong spin-orbit coupling and band splitting [4]. Of these, molybdenum disulfide (MoS_2) has received tremendous attention because of its unique crystal structure, and physical, chemical, and electrical properties [5–12]. Due to these fascinating properties, MoS_2 has been used in numerous applications, including photodetection, gas-sensing, and hydrogen evolution [13–17]. Atomically thin tungsten disulfide (WS_2) has a similar structure to MoS_2, but exhibits stronger photoluminescence (PL) quantum yield at room temperature and larger spin-orbit coupling compared to MoS_2 [18–20], which makes it promising for applications in new optoelectronics and spintronics. Nevertheless, the research on WS_2 lags far behind that of MoS_2 and the development of simple and controllable strategies for preparing high-quality and large-area atomically thin WS_2 is still a big challenge, which has impeded further fabrication of optoelectronics/spintronics devices.

As reported, large-area continuous mono- and multi-layer WS_2 were synthesized by sulfurizing the pre-deposited ultrathin tungsten or WO_x films obtained by various routes, including thermal evaporation, magnetron sputtering, and atomic layer deposition (ALD) [21–27]. This two-step chemical vapor deposition (CVD) method can provide large-area polycrystalline WS_2 film for the fabrication of devices. However, the size of the single-crystal domain in the polycrystalline film is limited to the nanoscale and its electrical properties could be seriously degraded because of the existence of abundant grain boundaries (GBs) [28]. Moreover, the pre-deposition of ultrathin films is time-consuming and more efforts should be paid to precisely control the thickness, which restricts the wide use of two-step CVD in research communities. To simplify the synthesis process, one-step CVD that is based on direct sulfurization of powdered WO_3 on diverse substrates, such as SiO_2/Si, sapphire, Au foils, and *h*-BN, has been employed for the synthesis of WS_2 [29–34]. In this direction, powdered sulfur and WO_3 in quartz/ceramic boats are placed at separate locations; normally the sulfur is located upstream and the WO_3 is placed underneath or ahead of the substrates. In addition, growth parameters, such as the temperature, heating rate, gas flow, and the distance between the precursor and substrate, can be accurately controlled between batches. However, this strategy presents some drawbacks since the distribution of the solid-phase precursors loaded in quartz/ceramic boats is not easy to control, and this could significantly influence the concentration of gaseous species on the growth interface. Furthermore, due to the difference in melting points of sulfur and WO_3, it is difficult to simultaneously and precisely control the evaporation of these two precursors and the subsequent transportation to the growth interface. As a result, the dimension and morphology of the CVD-grown WS_2 always present significant differences even under the same growth parameters.

Lee et al. [35] reported that continuous polycrystalline monolayer WS_2 films in the centimeter-scale were grown on SiO_2/Si substrate with the seeding of perylene-3,4,9,10-tetracarboxylic acid tetrapotassium salt (PTAS). However, the size of the isolated single-crystal domains was limited in the range of 10 to 20 μm. Zhang et al. [31] reported the synthesis of mono- and multi-layer WS_2 flakes with the domain size exceeding 50×50 μm^2 on the sapphire substrate at low-pressure with mixed hydrogen and argon gases. Yue et al. [36] reported that uniform triangular monolayer WS_2 flakes with a side length of ~233 μm were synthesized on SiO_2/Si substrate by carefully adjusting the introduction time of the sulfur precursor and the distance between the sources and substrates. Recently, Zhou et al. [37] demonstrated that molten-salt-assisted CVD can be used to synthesize large-area triangular monolayer WS_2 flakes with an edge length of 300 μm at moderate temperatures due to the reduction of the WO_3 melting point by using sodium chloride. To this end, significant attempts have been made towards the large-area synthesis of monolayer WS_2 by one-step CVD. However, the layer controllability and universality of the synthesis method are the major existing obstacles that prevent practical applications. Therefore, controllable synthesis of large-area monolayer WS_2 is still challenging.

In this study, we report a facile and controllable method for synthesis of large-area monolayer WS_2 flakes by one-step CVD at atmospheric pressure. To promote the distribution of WO_3, we first dispersed the powdered WO_3 in ethanol to form a suspension solution, then we used a pipette to drop-cast the solution onto the SiO_2/Si substrates. Moreover, we simultaneously loaded the powdered precursors and SiO_2/Si substrates in a small one-end sealed quartz tube; and we placed the small tube inside a bigger quartz tube to ensure adequate amounts of sulfur-precursors to participate in the whole process of WS_2 growth. Through this method, we obtained a series of triangular monolayer WS_2 flakes with edge lengths up to 250 to 370 μm and homogeneous crystallinity. The morphology, thickness, atomic structure, and light emissions of the as-grown WS_2 samples were characterized by various tools, such as an optical microscope (OM), atomic force microscope (AFM), transmission electron microscope (TEM), fluorescence (FL) microscope, PL spectrometer, and Raman spectrometer. It was found that the as-grown monolayer WS_2 flakes with edge lengths greater than 95 μm show suppressed fluorescence emission in the inner region, while the smaller one presents homogeneous fluorescence emission. Raman mapping results indicate that the distinctly size-dependent fluorescence emission is attributed to the heterogeneous release of intrinsic tensile strain within WS_2 after growth. Furthermore, the results of five independent experiments are consistent with each other, confirming the validity and reproducibility of this method.

2. Materials and Methods

In this research, the synthesis of WS_2 was achieved by one-step CVD in a horizontal single-zone furnace (Hefei FACEROM Co., Ltd., Hefei, China) at atmospheric pressure, as shown in Figure 1a. The CVD system mainly consists of the heating zone and a quartz tube with a 60 mm diameter. Upstream of the quartz tube is connected to the high-purity (99.999%) argon cylinder, while downstream is connected to the exhausted gas treatment system. We did not use hydrogen in this research, due to considerations of safety, although previous reports demonstrated that hydrogen could facilitate the reduction of WO_3 and lead to the growth of high quality WS_2 [38,39]. Furthermore, a small quartz tube with a diameter of 15 mm sealed at one-end was intentionally employed for holding precursors and substrates simultaneously, this procedure was different from previous reports [36,40,41], aiming at obtaining large-area WS_2 in a short period of time.

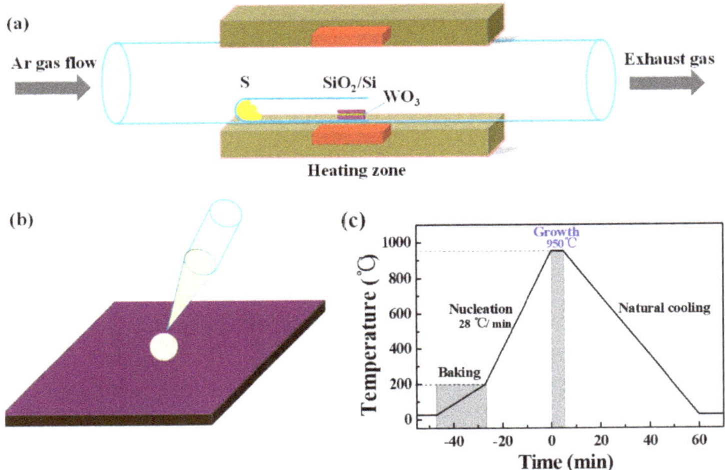

Figure 1. (**a**) Schematic diagram of the horizontal single-zone furnace employed for the synthesis of WS_2 on SiO_2/Si substrates. (**b**) Schematic illustration for the drop-casting of WO_3-ethanol suspension solution onto the SiO_2/Si substrate. (**c**) Temperature profile adopted for the synthesis of WS_2 flakes at 950 °C for 5 min.

The SiO_2 (300 nm)/Si (5 mm × 5 mm) sliced from a 4-inch wafer was used as substrate and was sequentially washed in DI (Deionized) water, acetone, DI water, ethanol, and DI water in an ultrasonic bath for 10 min each. The residual water and organic solvent were removed by compressed UHP (Ultra-high Purity) nitrogen gas blowing. Alcoholic tungsten suspension solution with a concentration of 1.5 mg/mL was prepared by WO_3 (>99.9%, Adamas-beta, Shanghai, China) and ethanol (AR, KESHI, Chengdu, China). Before the growth process, a drop (10 μL) of alcoholic tungsten suspension solution was directly dropped onto the cleaned substrate using a pipette, as shown in Figure 1b, and the substrate was dried on a heater (40 °C) for 3 min. A relatively uniform distribution of powdered WO_3 on the SiO_2/Si substrate was obtained (as shown in Supplementary Materials Figure S1) after the evaporation of ethanol. Then, another fresh and cleaned SiO_2/Si substrate was placed face-down above the drop-casted substrate to form a sandwich-structure. Subsequently, 600 mg of sulfur powder

(S sublimed, >99.5%, KESHI, Chengdu, China) was introduced at the bottom of the sealed end of the small quartz tube, while the sandwich-structure substrates were carefully placed downstream, 24 cm away from the powdered sulfur. After that, the small quartz tube together with the precursors and substrates was loaded into the bigger quartz tube carefully, ensuring the substrates were exactly located at the center of the heating zone. Before heating, the furnace chamber was pumped to low pressure (<10 Pa) and then purged by 200 sccm argon to atmospheric pressure. Figure 1c shows the temperature profile used in this study. Initially, the furnace was heated from room temperature to 200 °C with a heating rate of 10 °C/min to remove the contaminants, such as water or residual organics. Subsequently, a higher heating rate of 28 °C/min was used to increase the temperature to 950 °C to obtain a high nucleation rate of WS_2. The growth of WS_2 was kept at 950 °C for 5 min under 200 sccm of argon. After growth, the furnace was cooled naturally to room temperature.

The morphology and size of the as-grown WS_2 samples were characterized by an OM (50i POL, Nikon, Tokyo, Japan). The thickness and atomic structure of the as-grown WS_2 were analyzed via an AFM (Dimension EDGE, Bruker, Billerica, MA, USA) in tapping mode and TEM (Tecnai G^2 F20, FEI, Hillsboro, OR, USA), respectively. Raman spectra were collected in the backscattering geometry at room temperature with an excitation wavelength of 532 nm (InVia Reflex, Renishaw, Gloucestershire, UK). Before Raman characterization, the system was calibrated with the Raman peak of Si at 520 cm^{-1}. Photoluminescence spectra were obtained from a home-made laser scanning confocal microscope system. A 532 nm CW laser with an average power of 100 μW was employed as the excitation source to avoid sample damage. Fluorescence images were captured by using a fluorescence microscope (BX53, Olympus, Tokyo, Japan) under excitation of a green light source.

3. Results and Discussion

Figure 2a shows a typical optical image of the as-grown WS_2 sample on the covering SiO_2/Si substrate, which is blank without WO_3-ethanol drop-casting. The as-grown WS_2 exhibits the feature of an equilateral triangle with sharp edges. Furthermore, the GBs can also be observed, which is caused by the coalescence of adjacent triangles and is hard to avoid in CVD-grown MoS_2 and WS_2 flakes [6,28]. It is noted that all the triangles regardless of their sizes show uniform contrast under OM, indicating good uniformity of the thickness of the as-grown WS_2. Figure 2b presents a representative optical image of the WS_2 sample grown on the bottom SiO_2/Si substrate, which was drop-casted by WO_3-ethanol solution. In contrast with the covering substrate, small triangular WS_2 with uniform contrast can be occasionally observed, however, irregular-shape WS_2 with heterogeneous contrast and particulate WS_2 (as shown in Supplementary Materials Figure S2) are predominant on the bottom substrate. These distinctly morphological features could be caused by the existence of the powdered WO_3 and highly concentrated reaction sources on the bottom substrate [33]. Due to the differences in the nucleation density and local environment on the substrate, the dimension distribution of the triangular flakes is quite wide, from tens to hundreds of micrometers, while large triangular WS_2 flakes with edge lengths of ~270 μm are readily obtained on the covering SiO_2/Si substrate, as shown in Figure 2a. In addition, truncated-triangle, hexagon, and butterfly-shaped flakes are formed on the covering SiO_2/Si substrate as well (as shown in Supplementary Materials Figure S3), which was observed in previous CVD-grown WS_2 and MoS_2 and explained from the point of the changes in the M:S (M = Mo, W) ratio of the precursors within the growth interface [6,31,42,43].

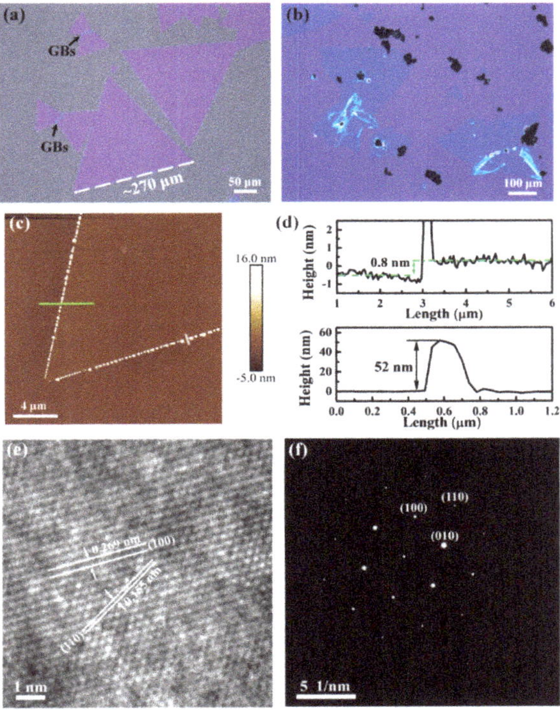

Figure 2. Representative optical images of the as-grown WS_2 flakes on the covering (**a**) and bottom (**b**) SiO_2/Si substrates, respectively. AFM image (**c**) and its corresponding height profiles (**d**) of an equilateral triangular WS_2 flake on SiO_2/Si substrate. High-resolution TEM image (**e**) and its corresponding SAED pattern (**f**) of the as-grown triangular WS_2 transferred on a holy carbon-coated copper TEM grid.

In order to accurately determine the thickness of the as-grown WS_2 sample, AFM was performed in tapping mode. Figure 2c,d show the AFM image and its corresponding height profiles of an equilateral triangular WS_2 flake with uniform contrast under OM. It can be clearly observed that there are numerous small particles with several hundreds of nanometers in length and dozens of nanometers in height along the triangle edges. This phenomenon was mentioned by a previous report [37], and it is probably caused by the continuous attachment of the precursors to the growing edge. The thickness of the as-grown WS_2 flake determined by the height profile is ~0.8 nm, which is consistent with the reported thickness of monolayer WS_2 [36,44], corroborating the monolayer nature of the triangular WS_2 flake with uniform contrast obtained under OM. To further identify the atomic structure of our sample, the triangular WS_2 was transferred to a holy carbon-coated copper grid via the wet transfer method and analyzed by TEM. Figure 2e shows the high resolution TEM image of our WS_2 sample, which clearly shows the hexagonal ring lattice consisting of alternating tungsten and sulfur atoms. Furthermore, the corresponding selective area electron diffraction (SAED) displayed in Figure 2f reveals only one set of diffraction spots, demonstrating the single-crystal nature of our WS_2 sample. The inter-planar distances of (100) and (110) planes deduced from high resolution TEM measurement are ~0.269 nm and 0.155 nm, respectively, which coincide with the previous reported results of CVD-grown monolayer WS_2 [40].

As a powerful and nondestructive tool, the Raman spectrometer has been widely employed to study the properties of TMDs, such as determination of the layer number [45–47], electrostatic doping [48], assessment of crystallinity [49,50], as well as internal and external strain [51,52]. Figure 3a shows a typical Raman spectrum of an equilateral triangular WS_2 flake collected in the backscattering geometry at room temperature. As shown in Figure 3a, except the peak at 520 cm^{-1} from Si substrate, the other peaks are attributed to the WS_2 flake. The strongest peak at ~350 cm^{-1} can be fitted well with three sub-peaks with a maximum frequency located at 344.7, 350.0, and 354.6 cm^{-1}, and they can be identified as the first-order optical mode of $E^1_{2g}(M)$, the second-order longitudinal acoustic mode of 2LA (M), and the first-order optical mode of $E^1_{2g}(\Gamma)$ (in-plane vibration between sulfur and tungsten atoms as shown in the inset of Figure 3a), respectively, according to the theoretical and experimental studies [45,53]. In addition, strong combination modes of 2LA (M) $-$ $2E^2_{2g}(\Gamma)$ and 2LA (M) $-$ $E^2_{2g}(\Gamma)$ were observed at 294.8 and 321.7 cm^{-1}, respectively. The appearance of these strong combination modes is mainly attributed to the strong resonance between the phonon and B-exciton in WS_2 excited by the 532 nm laser. The peak located at 416.8 cm^{-1} is assigned to the $A_{1g}(\Gamma)$ mode, which is caused by the out-of-plane vibration between sulfur atoms and is sensitive to the layer number of WS_2 [46]. Moreover, the intensity of 2LA (M) is much stronger than that of the $A_{1g}(\Gamma)$ mode, giving rise to an intensity ratio ($R = I_{2LA\ (M)}/I_{A_{1g}(\Gamma)}$) of ~5.6, and the difference between the frequencies of the $A_{1g}(\Gamma)$ and $E^1_{2g}(\Gamma)$ modes ($\Delta = A_{1g}(\Gamma) - E^1_{2g}(\Gamma)$) is ~62.2 cm^{-1}. These characteristics are quite consistent with the reported results for monolayer WS_2 excited by the 532 nm laser [33,36,41]. To probe the light emission and further determine the layer number of the triangular WS_2 flake, the room temperature PL spectrum was collected with a 532 nm laser excitation. As shown in Figure 3b, a single and strong peak with a maxima wavelength of 632 nm (~1.96 eV) is observed, which is consistent with the reported PL peak position for monolayer WS_2 [41,54,55], again confirming the monolayer nature of the as-grown WS_2.

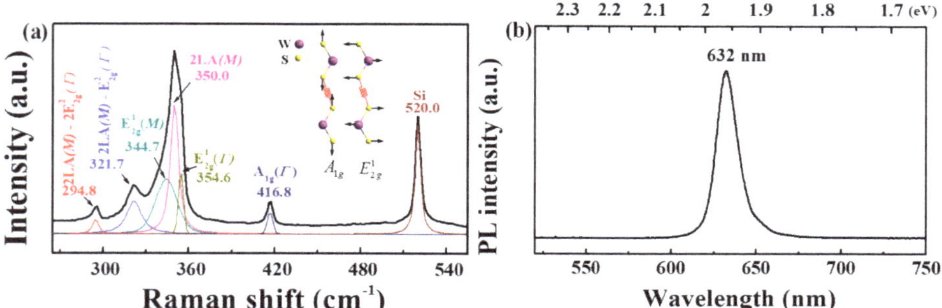

Figure 3. Room temperature Raman (**a**) and PL (**b**) spectra collected from an equilateral triangular WS_2 flake on SiO_2/Si substrate with 532 nm excitation. The inset plotted in Figure 3a shows the schematic diagram of the atomic vibrations of $A_{1g}(\Gamma)$ and $E^1_{2g}(\Gamma)$ modes in WS_2. It is noted that the black curve is raw data while the color ones are obtained by Lorentz fitting as shown in Figure 3a.

To investigate the light emission uniformity of the as-grown triangular WS_2 flake, FL image was captured under excitation of a green light source. Figure 4a clearly shows that the outer region of the triangular WS_2 flake with an edge length of ~270 μm exhibits intense FL emission, and it becomes weaker towards the inner region. In contrast, the two coalescent small triangles show relatively uniform FL emission across the entire region. This inhomogeneous FL emission in CVD-grown monolayer WS_2 was observed by other groups. Peimyoo et al. [18] reported a similar observation of the suppression of FL emission for the center region as compared with the edge and they speculated that the suppression was relative to the structural imperfection and n-doping induced by charged defects. Liu et al. [56] observed inhomogeneous FL patterns consisting of alternating dark and bright concentric triangles in monolayer WS_2, and they attributed the darker region to the high concentration of sulfur vacancies.

Recently, Feng et al. [57] reported a novel FL aging behavior in monolayer WS_2 with a large size, and they attributed that behavior to the partial release of intrinsic tensile strain after CVD growth.

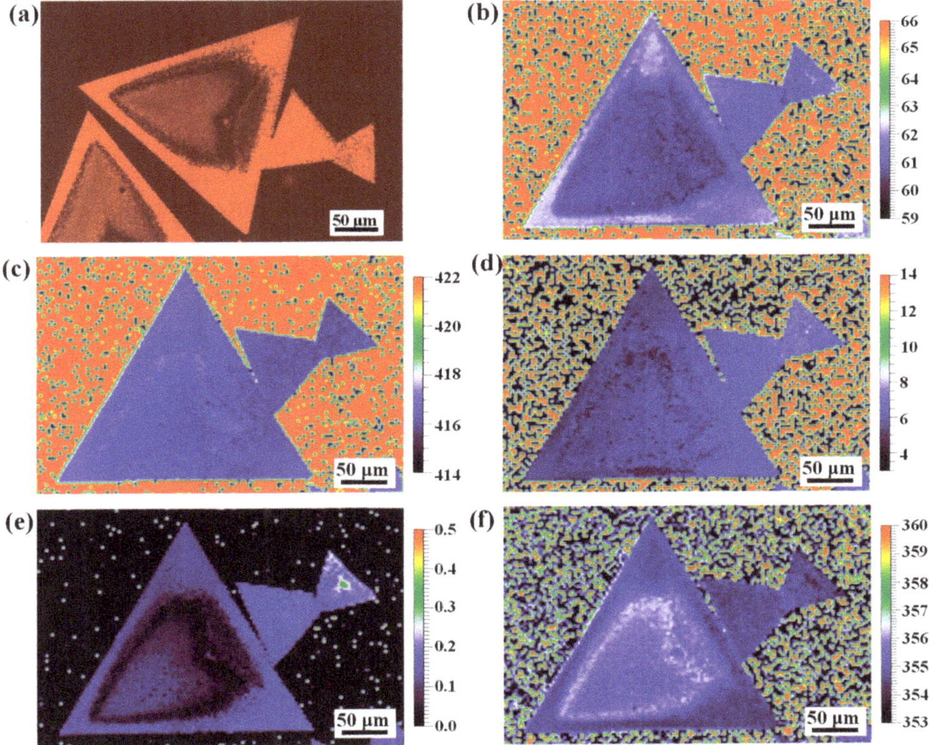

Figure 4. (**a**) FL image of the as-grown WS_2 flake on SiO_2/Si substrate, its optical image is shown in Figure 2a. (**b–f**) Raman mappings of the specific WS_2 sample: frequency difference between $A_{1g}(\Gamma)$ and $E^1_{2g}(\Gamma)$ modes (**b**), frequency of $A_{1g}(\Gamma)$ mode (**c**), full width of half maximum (FWHM) of $A_{1g}(\Gamma)$ mode (**d**), normalized intensity of $A_{1g}(\Gamma)$ mode (**e**), and frequency of $E^1_{2g}(\Gamma)$ mode (**f**).

To better understand the heterogeneity of FL emission within the large-area triangular WS_2, Raman mapping with a step of 3 µm was performed and more than 10,000 data points across the entire flake were collected. As shown in Figure 4b, the distribution of the frequency difference between the $A_{1g}(\Gamma)$ and $E^1_{2g}(\Gamma)$ modes is in the range of 61 to 62.5 cm^{-1}, which is accordance with the reported results for monolayer WS_2 [33,36,45], demonstrating that the entire triangular WS_2 possesses a monolayer nature. It has been reported that the $A_{1g}(\Gamma)$ mode is not only sensitive to the layer number, but also to the electrostatic doping in the typical 2*H*-type TMDs, and it will red-shift and broaden with electron doping [48]. As shown in Figure 4c,d, the frequency and full width of half maximum (FWHM) mappings of the $A_{1g}(\Gamma)$ mode show relatively uniform features, confirming good uniformity of the crystallinity and electron doping for this large WS_2 triangle. As a result, the heterogeneous FL emission observed in this WS_2 triangle is not related to the n-doping. In addition, it has been reported that the frequency of the $E^1_{2g}(\Gamma)$ mode will blue-shift and the intensity of the $A_{1g}(\Gamma)$ mode will decrease after release of the tensile strain in monolayer WS_2 [51,57]. To investigate the stress distribution, the normalized intensity mapping of the $A_{1g}(\Gamma)$ mode and frequency mapping of the $E^1_{2g}(\Gamma)$ mode for this WS_2 triangle were analyzed, as shown in Figure 4e,f, respectively. It can be clearly seen that the intensity of the $A_{1g}(\Gamma)$ mode decreases obviously and the frequency of the $E^1_{2g}(\Gamma)$ mode blue-shifts

~1 cm^{-1} for the inner region as compared with the outer region. This indicates that the tensile strain in the outer region is stronger than that of the inner region. By carefully comparing the FL image with the Raman mappings of the $A_{1g}(\Gamma)$ mode intensity and $E^1_{2g}(\Gamma)$ mode frequency, it can be seen that they match each other well. Therefore, the heterogeneity of FL emission observed in the large WS_2 triangle is mainly ascribed to the inhomogeneous tensile strain in the WS_2.

Figure 5 illustrates the FL images of the as-grown WS_2 flakes with different edge lengths. The FL image of the WS_2 triangle with an edge length of ~171 μm displayed in Figure 5a possesses similar heterogeneity of the FL emission as the larger WS_2 triangle observed in Figure 4a. Furthermore, it can be seen that the red large triangle is divided into three smaller and equilateral triangles by three suppressed lines, which is similar to the previous observation in a WS_2 triangle with an edge size ~100 μm and disappears with aging time [57]. The optical image and Raman mappings of this WS_2 triangle are shown in Supplementary Materials Figure S4. It can be seen that the Raman mapping results for this WS_2 triangle present similar features to the larger one observed in Figure 4a, so that the heterogeneity of the FL emission observed in this WS_2 triangle is ascribed to the inhomogeneous tensile strain as well. When the edge length of the WS_2 triangle is reduced to ~95 μm, three lines with slightly weak FL emission can also be observed in the red triangle displayed in Figure 5b, whereas the FL emission shows better homogeneity across the entire triangle. Interestingly, the FL emission becomes more homogeneous and intense when the edge length of the WS_2 triangle is less than ~32 μm, as shown in Figure 5c. Therefore, it is evident that the monolayer triangular WS_2 flakes synthesized in this study present distinctly size-dependent FL emission.

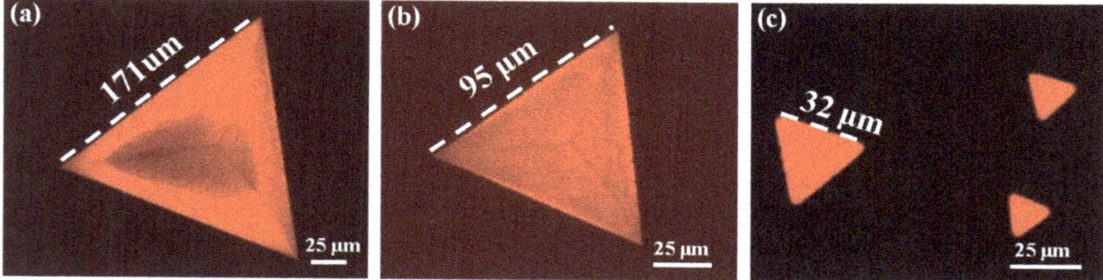

Figure 5. FL images of the as-grown WS_2 flakes with different edge lengths: ~171 μm (**a**), ~95 μm (**b**), and less than 32 μm (**c**).

Intrinsic tensile strain could be introduced in monolayer MS_2 grown on SiO_2/Si substrate during the fast cooling process from the high growth temperature to room temperature due to the higher thermal expansion coefficient of MS_2 compared with that of silica substrate [57,58]. Feng et al. [57] reported that the intrinsic tensile strain could be partially released from the edge towards the center with the aging time, and consequently concentric FL patterns could be formed in large size (~100 μm) monolayer WS_2 crystals after 72 h of aging. In this study, all the FL images were captured a week after synthesis, however, we did not detect the concentric FL patterns, probably because the release of intrinsic tensile strain was related to the geometry of the WS_2 flake and the interaction between WS_2 and the substrates.

To evaluate the validity and reproducibility of this method, we performed an extra four batches of experiments under the same conditions (950 °C for 5 min). Figure 6 shows the representative optical images of the as-grown WS_2 flakes obtained from different bathes of experiments under the same conditions. We clearly noted that triangular WS_2 flakes with an edge length of 250 to 370 μm were readily formed for each batch of the experiment, indicating that the distribution of the solid-sate precursor was controlled well by the drop-casting method. Recently, centimeter scale continuous WS_2 films with GBs and triangular monolayer WS_2 flakes hundreds of micrometers (~300 μm) in dimension were grown by direct sulfurization of powdered WO_3 on SiO_2/Si substrates, however, the growth time for previous studies was more than 10 min or even more [28,36,57]. In this study, triangular monolayer

WS_2 flakes with an edge length of 250 to 370 μm were effectively formed only within 5 min CVD growth. Therefore, the transportation of precursors could be limited and thus more precursors were able to participate in the synthesis of WS_2 when a small one-end sealed quartz tube was employed for holding the precursors and substrates simultaneously.

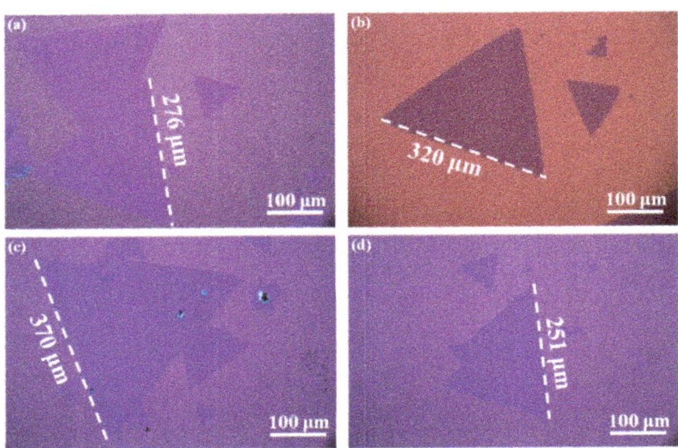

Figure 6. (**a–d**) Representative optical images of the as-grown WS_2 flakes on the covering SiO_2/Si substrates obtained from different batches of experiments under the same conditions (950 °C for 5 min.).

4. Conclusions

We demonstrated a facile and controllable method for the synthesis of large-area monolayer WS_2 flakes by direct sulfurization of powdered WO_3 on SiO_2/Si substrates in a horizontal single-zone CVD system. A series of monolayer WS_2 flakes with an edge length of 250 to 370 μm on SiO_2/Si substrates were achieved by five independent growth experiments. The morphology, thickness, atomic structure, and light emission of the as-grown WS_2 samples were characterized using various tools. It was found that the as-grown monolayer WS_2 flakes exhibit homogeneous crystallinity and size-dependent FL emission. For the WS_2, triangles with edge lengths less than 95 μm exhibited uniform FL emission while that with larger edge lengths showed heterogeneous FL emission. This heterogeneity of FL emission in large monolayer WS_2 flakes can be attributed to the heterogeneous release of intrinsic tensile strain after growth. We believe our method could offer an opportunity that opens a new window for the growth of other TMDCs with large areas.

Supplementary Materials: The following are available online at http://www.mdpi.com/2079-4991/9/4/578/s1, Figure S1: Optical image of powdered WO_3 on SiO_2/Si substrate after the evaporation of ethanol. Figure S2: Raman spectra of the WO_3 powder (a), sulfur powder (b) and particulate WS_2 (c) marked with white arrow in Figure S2(d). Representative optical image of the as-grown WS_2 sample on the bottom SiO_2/Si substrate that was drop-casted by WO_3-ethanol solution (d). Figure S3: Optical images of as-grown WS_2 flakes on SiO_2/Si substrate with various morphologies: Truncated-triangle (a), Hexagon (b, Butterfly (c). Figure S4: Optical image of the WS_2 flake with edge length of ~171 μm on SiO_2/Si substrate (a). Raman mappings of the specific WS_2 sample: frequency difference between $A_{1g}(\Gamma)$ and $E^1_{2g}(\Gamma)$ modes (b); frequency (c), full width of half maximum (FWHM) (d) and normalized intensity (e) of $A_{1g}(\Gamma)$ mode; frequency of $E^1_{2g}(\Gamma)$ mode (f).

Author Contributions: Conceptualization, B.S. and D.W.; Data curation, K.D.; Formal analysis, S.F. (Shaoxi Fang), S.F. (Shuanglong Feng) and H.Z.; Funding acquisition, D.Z.; Investigation, B.S. and D.Z.; Project administration, D.W.; Supervision, C.T. and D.W.; Writing—original draft, B.S.; Writing—review & editing, D.Z. and C.T. B.S. and D.Z. contributed equally to this work.

Funding: This work was supported by National Natural Science Foundation of China (Grant No. 61701474), Natural Science Foundation of Chongqing, China (Grant No. cstc2017jcyjAX0320) and West Light Foundation of CAS.

Acknowledgments: The authors would like to thank Yuzhi Li for PL measurements, Risheng Qiu for TEM measurements and Xilin Tan for polishing writing.

Conflicts of Interest: The authors declare no conflicts of interest.

References

1. Ji, Q.; Zhang, Y.; Zhang, Y.; Liu, Z. Chemical vapour deposition of group-VIB metal dichalcogenide monolayers: Engineered substrates from amorphous to single crystalline. *Chem. Soc. Rev.* **2015**, *44*, 2587–2602. [CrossRef] [PubMed]
2. Shi, J.P.; Ji, Q.Q.; Liu, Z.F.; Zhang, Y.F. Recent advances in controlling syntheses and energy related applications of MX_2 and MX_2/graphene heterostructures. *Adv. Energy Mater.* **2016**, *6*, 24. [CrossRef]
3. Brent, J.R.; Savjani, N.; O'Brien, P. Synthetic approaches to two-dimensional transition metal dichalcogenide nanosheets. *Prog. Mater. Sci.* **2017**, *89*, 411–478. [CrossRef]
4. Wang, Q.H.; Kalantar-Zadeh, K.; Kis, A.; Coleman, J.N.; Strano, M.S. Electronics and optoelectronics of two-dimensional transition metal dichalcogenides. *Nat. Nanotechnol.* **2012**, *7*, 699–712. [CrossRef] [PubMed]
5. Najmaei, S.; Liu, Z.; Zhou, W.; Zou, X.; Shi, G.; Lei, S.; Yakobson, B.I.; Idrobo, J.-C.; Ajayan, P.M.; Lou, J. Vapour phase growth and grain boundary structure of molybdenum disulphide atomic layers. *Nat. Mater.* **2013**, *12*, 754–759. [CrossRef] [PubMed]
6. van der Zande, A.M.; Huang, P.Y.; Chenet, D.A.; Berkelbach, T.C.; You, Y.; Lee, G.-H.; Heinz, T.F.; Reichman, D.R.; Muller, D.A.; Hone, J.C. Grains and grain boundaries in highly crystalline monolayer molybdenum disulphide. *Nat. Mater.* **2013**, *12*, 554–561. [CrossRef] [PubMed]
7. Cai, L.; He, J.; Liu, Q.; Yao, T.; Chen, L.; Yan, W.; Hu, F.; Jiang, Y.; Zhao, Y.; Hu, T.; et al. Vacancy-induced ferromagnetism of MoS_2 nanosheets. *J. Am. Chem. Soc.* **2015**, *137*, 2622–2627. [CrossRef] [PubMed]
8. Taube, A.; Judek, J.; Jastrzebski, C.; Duzynska, A.; Switkowski, K.; Zdrojek, M. Temperature-dependent nonlinear phonon shifts in a supported MoS_2 monolayer. *ACS Appl. Mater. Interfaces* **2014**, *6*, 8959–8963. [CrossRef] [PubMed]
9. Lanzillo, N.A.; Birdwell, A.G.; Amani, M.; Crowne, F.J.; Shah, P.B.; Najmaei, S.; Liu, Z.; Ajayan, P.M.; Lou, J.; Dubey, M.; et al. Temperature-dependent phonon shifts in monolayer MoS_2. *Appl. Phys. Lett.* **2013**, *103*, 093102. [CrossRef]
10. Splendiani, A.; Sun, L.; Zhang, Y.; Li, T.; Kim, J.; Chim, C.-Y.; Galli, G.; Wang, F. Emerging photoluminescence in monolayer MoS_2. *Nano Lett.* **2010**, *10*, 1271–1275. [CrossRef]
11. Zhao, H.-Q.; Mao, X.; Zhou, D.; Feng, S.; Shi, X.; Ma, Y.; Wei, X.; Mao, Y. Bandgap modulation of MoS_2 monolayer by thermal annealing and quick cooling. *Nanoscale* **2016**, *8*, 18995–19003. [CrossRef] [PubMed]
12. Radisavljevic, B.; Radenovic, A.; Brivio, J.; Giacometti, V.; Kis, A. Single-layer MoS_2 transistors. *Nat. Nanotechnol.* **2011**, *6*, 147–150. [CrossRef] [PubMed]
13. Voiry, D.; Salehi, M.; Silva, R.; Fujita, T.; Chen, M.; Asefa, T.; Shenoy, V.B.; Eda, G.; Chhowalla, M. Conducting MoS_2 Nanosheets as Catalysts for Hydrogen evolution reaction. *Nano Lett.* **2013**, *13*, 6222–6227. [CrossRef] [PubMed]
14. Lopez-Sanchez, O.; Lembke, D.; Kayci, M.; Radenovic, A.; Kis, A. Ultrasensitive photodetectors based on monolayer MoS_2. *Nat. Nanotechnol.* **2013**, *8*, 497–501. [CrossRef]
15. Zhang, W.; Huang, J.-K.; Chen, C.-H.; Chang, Y.-H.; Cheng, Y.-J.; Li, L.-J. High-gain phototransistors based on a CVD MoS_2 monolayer. *Adv. Mater.* **2013**, *25*, 3456–3461. [CrossRef] [PubMed]
16. Li, H.; Yin, Z.; He, Q.; Li, H.; Huang, X.; Lu, G.; Fam, D.W.H.; Tok, A.I.Y.; Zhang, Q.; Zhang, H. Fabrication of single- and multilayer MoS_2 film-based field-effect transistors for sensing NO at room temperature. *Small* **2012**, *8*, 63–67. [CrossRef] [PubMed]
17. Late, D.J.; Huang, Y.-K.; Liu, B.; Acharya, J.; Shirodkar, S.N.; Luo, J.; Yan, A.; Charles, D.; Waghmare, U.V.; Dravid, V.P.; et al. Sensing behavior of atomically thin-layered MoS_2 transistors. *ACS Nano* **2013**, *7*, 4879–4891. [CrossRef] [PubMed]
18. Peimyoo, N.; Shang, J.; Cong, C.; Shen, X.; Wu, X.; Yeow, E.K.L.; Yu, T. Nonblinking, intense two-dimensional light emitter: Mono layer WS_2 triangles. *ACS Nano* **2013**, *7*, 10985–10994. [CrossRef] [PubMed]
19. Zhao, W.; Ghorannevis, Z.; Chu, L.; Toh, M.; Kloc, C.; Tan, P.-H.; Eda, G. Evolution of electronic structure in atomically thin sheets of WS_2 and WSe_2. *ACS Nano* **2013**, *7*, 791–797. [CrossRef] [PubMed]
20. Zhu, Z.Y.; Cheng, Y.C.; Schwingenschloegl, U. Giant spin-orbit-induced spin splitting in two-dimensional transition-metal dichalcogenide semiconductors. *Phys. Rev. B* **2011**, *84*, 153402. [CrossRef]
21. Song, J.-G.; Park, J.; Lee, W.; Choi, T.; Jung, H.; Lee, C.W.; Hwang, S.-H.; Myoung, J.M.; Jung, J.-H.; Kim, S.-H.; et al. Layer-controlled, wafer-scale, and conformal synthesis of tungsten disulfide nanosheets using atomic layer deposition. *ACS Nano* **2013**, *7*, 11333–11340. [CrossRef] [PubMed]

22. Elias, A.L.; Perea-Lopez, N.; Castro-Beltran, A.; Berkdemir, A.; Lv, R.T.; Feng, S.M.; Long, A.D.; Hayashi, T.; Kim, Y.A.; Endo, M.; et al. Controlled synthesis and transfer of large-area WS_2 sheets: From single layer to few layers. *ACS Nano* **2013**, *7*, 5235–5242. [CrossRef]
23. Orofeo, C.M.; Suzuki, S.; Sekine, Y.; Hibino, H. Scalable synthesis of layer-controlled WS_2 and MoS_2 sheets by sulfurization of thin metal films. *Appl. Phys. Lett.* **2014**, 105. [CrossRef]
24. Jung, Y.; Shen, J.; Liu, Y.; Woods, J.M.; Sun, Y.; Cha, J.J. Metal Seed Layer Thickness-induced transition from vertical to horizontal growth of MoS_2 and WS_2. *Nano Lett.* **2014**, *14*, 6842–6849. [CrossRef] [PubMed]
25. Park, J.; Kim, M.S.; Cha, E.; Kim, J.; Choi, W. Synthesis of uniform single layer WS_2 for tunable photoluminescence. *Sci. Rep.* **2017**, *7*, 16121. [CrossRef]
26. Chen, Y.; Gan, L.; Li, H.; Ma, Y.; Zhai, T. Achieving uniform monolayer transition metal dichalcogenides film on silicon wafer via silanization treatment: A typical study on WS_2. *Adv. Mater.* **2017**, *29*, 1603550. [CrossRef] [PubMed]
27. Zhang, Y.S.; Shi, J.P.; Han, G.F.; Li, M.J.; Ji, Q.Q.; Ma, D.L.; Zhang, Y.; Li, C.; Lang, X.Y.; Zhang, Y.F.; et al. Chemical vapor deposition of monolayer WS_2 nanosheets on Au foils toward direct application in hydrogen evolution. *Nano Res.* **2015**, *8*, 2881–2890. [CrossRef]
28. Jia, Z.; Hu, W.; Xiang, J.; Wen, F.; Nie, A.; Mu, C.; Zhao, Z.; Xu, B.; Tian, Y.; Liu, Z. Grain wall boundaries in centimeter-scale continuous monolayer WS_2 film grown by chemical vapor deposition. *Nanotechnology* **2018**, *29*, 255705. [CrossRef] [PubMed]
29. Okada, M.; Sawazaki, T.; Watanabe, K.; Taniguch, T.; Hibino, H.; Shinohara, H.; Kitaura, R. Direct chemical vapor deposition growth of WS_2 atomic layers on hexagonal boron nitride. *ACS Nano* **2014**, *8*, 8273–8277. [CrossRef] [PubMed]
30. Xu, Z.Q.; Zhang, Y.P.; Lin, S.H.; Zheng, C.X.; Zhong, Y.L.; Xia, X.; Li, Z.P.; Sophia, P.J.; Fuhrer, M.S.; Cheng, Y.B.; et al. Synthesis and transfer of large-area monolayer WS_2 crystals: Moving toward the recyclable use of sapphire substrates. *ACS Nano* **2015**, *9*, 6178–6187. [CrossRef] [PubMed]
31. Zhang, Y.; Zhang, Y.F.; Ji, Q.Q.; Ju, J.; Yuan, H.T.; Shi, J.P.; Gao, T.; Ma, D.L.; Liu, M.X.; Chen, Y.B.; et al. Controlled growth of high-quality monolayer WS_2 layers on sapphire and imaging its grain boundary. *ACS Nano* **2013**, *7*, 8963–8971. [CrossRef] [PubMed]
32. Gao, Y.; Liu, Z.B.; Sun, D.M.; Huang, L.; Ma, L.P.; Yin, L.C.; Ma, T.; Zhang, Z.Y.; Ma, X.L.; Peng, L.M.; et al. Large-area synthesis of high-quality and uniform monolayer WS_2 on reusable Au foils. *Nat. Commun.* **2015**, *6*, 8569. [CrossRef]
33. Cong, C.X.; Shang, J.Z.; Wu, X.; Cao, B.C.; Peimyoo, N.; Qiu, C.; Sun, L.T.; Yu, T. Synthesis and optical properties of large-area single-crystalline 2D semiconductor WS_2 monolayer from chemical vapor deposition. *Adv. Opt. Mater.* **2014**, *2*, 131–136. [CrossRef]
34. Lan, F.; Yang, R.; Xu, Y.; Qian, S.; Zhang, S.; Cheng, H.; Zhang, Y. Synthesis of large-scale single-crystalline monolayer WS_2 using a semi-sealed Method. *Nanomaterials* **2018**, *8*, 100. [CrossRef] [PubMed]
35. Lee, Y.H.; Yu, L.L.; Wang, H.; Fang, W.J.; Ling, X.; Shi, Y.M.; Lin, C.T.; Huang, J.K.; Chang, M.T.; Chang, C.S.; et al. Synthesis and transfer of single-layer transition metal disulfides on diverse surfaces. *Nano Lett.* **2013**, *13*, 1852–1857. [CrossRef] [PubMed]
36. Yue, Y.; Chen, J.; Zhang, Y.; Ding, S.; Zhao, F.; Wang, Y.; Zhang, D.; Li, R.; Dong, H.; Hu, W.; et al. Two-dimensional high-quality monolayered triangular WS_2 flakes for field-effect transistors. *ACS Appl. Mater. Interfaces* **2018**, *10*, 22435–22444. [CrossRef] [PubMed]
37. Zhou, J.; Lin, J.; Huang, X.; Zhou, Y.; Chen, Y.; Xia, J.; Wang, H.; Xie, Y.; Yu, H.; Lei, J.; et al. A library of atomically thin metal chalcogenides. *Nature* **2018**, *556*, 355–361. [CrossRef] [PubMed]
38. McCreary, K.M.; Hanbicki, A.T.; Jernigan, G.G.; Culbertson, J.C.; Jonker, B.T. Synthesis of large-area WS_2 monolayers with exceptional photoluminescence. *Sci. Rep.* **2016**, *6*, 19159. [CrossRef] [PubMed]
39. Fu, Q.; Wang, W.H.; Yang, L.; Huang, J.; Zhang, J.Y.; Xiang, B. Controllable synthesis of high quality monolayer WS_2 on a SiO_2/Si substrate by chemical vapor deposition. *RSC Adv.* **2015**, *5*, 15795–15799. [CrossRef]
40. Liu, P.; Luo, T.; Xing, J.; Xu, H.; Hao, H.; Liu, H.; Dong, J. Large-area WS_2 film with big single domains grown by chemical vapor deposition. *Nanoscale Res. Lett.* **2017**, *12*, 558. [CrossRef]
41. Rong, Y.M.; Fan, Y.; Koh, A.L.; Robertson, A.W.; He, K.; Wang, S.S.; Tan, H.J.; Sinclair, R.; Warner, J.H. Controlling sulphur precursor addition for large single crystal domains of WS_2. *Nanoscale* **2014**, *6*, 12096–12103. [CrossRef] [PubMed]

42. Wang, S.; Rong, Y.; Fan, Y.; Pacios, M.; Bhaskaran, H.; He, K.; Warner, J.H. Shape evolution of monolayer MoS_2 crystals grown by chemical vapor deposition. *Chem. Mater.* **2014**, *26*, 6371–6379. [CrossRef]
43. Thangaraja, A.; Shinde, S.M.; Kalita, G.; Tanemura, M. An effective approach to synthesize monolayer tungsten disulphide crystals using tungsten halide precursor. *Appl. Phys. Lett.* **2016**, *108*, 053104. [CrossRef]
44. McCreary, K.M.; Hanbicki, A.T.; Singh, S.; Kawakami, R.K.; Jernigan, G.G.; Ishigami, M.; Ng, A.; Brintlinger, T.H.; Stroud, R.M.; Jonker, B.T. The Effect of preparation conditions on Raman and photoluminescence of monolayer WS_2. *Sci. Rep.* **2016**, *6*, 35154. [CrossRef] [PubMed]
45. Berkdemir, A.; Gutierrez, H.R.; Botello-Mendez, A.R.; Perea-Lopez, N.; Elias, A.L.; Chia, C.I.; Wang, B.; Crespi, V.H.; Lopez-Urias, F.; Charlier, J.C.; et al. Identification of individual and few layers of WS_2 using Raman Spectroscopy. *Sci. Rep.* **2013**, *3*, 1755. [CrossRef]
46. Zhao, W.; Ghorannevis, Z.; Amara, K.K.; Pang, J.R.; Toh, M.; Zhang, X.; Kloc, C.; Tan, P.H.; Eda, G. Lattice dynamics in mono- and few-layer sheets of WS_2 and WSe_2. *Nanoscale* **2013**, *5*, 9677–9683. [CrossRef] [PubMed]
47. Lee, C.; Yan, H.; Brus, L.E.; Heinz, T.F.; Hone, J.; Ryu, S. Anomalous lattice vibrations of single- and few-layer MoS_2. *ACS Nano* **2010**, *4*, 2695–2700. [CrossRef] [PubMed]
48. Chakraborty, B.; Bera, A.; Muthu, D.V.S.; Bhowmick, S.; Waghmare, U.V.; Sood, A.K. Symmetry-dependent phonon renormalization in monolayer MoS_2 transistor. *Phys. Rev. B* **2012**, *85*, 161403. [CrossRef]
49. McCreary, A.; Berkdemir, A.; Wang, J.J.; Nguyen, M.A.; Elias, A.L.; Perea-Lopez, N.; Fujisawa, K.; Kabius, B.; Carozo, V.; Cullen, D.A.; et al. Distinct photoluminescence and Raman spectroscopy signatures for identifying highly crystalline WS_2 monolayers produced by different growth methods. *J. Mater. Res.* **2016**, *31*, 931–944. [CrossRef]
50. Mignuzzi, S.; Pollard, A.J.; Bonini, N.; Brennan, B.; Gilmore, I.S.; Pimenta, M.A.; Richards, D.; Roy, D. Effect of disorder on Raman scattering of single-layer MoS_2. *Phys. Rev. B* **2015**, *91*, 195411. [CrossRef]
51. Wang, Y.; Cong, C.; Yang, W.; Shang, J.; Peimyoo, N.; Chen, Y.; Kang, J.; Wang, J.; Huang, W.; Yu, T. Strain-induced direct-indirect bandgap transition and phonon modulation in monolayer WS_2. *Nano Res.* **2015**, *8*, 2562–2572. [CrossRef]
52. Wang, Y.; Cong, C.; Qiu, C.; Yu, T. Raman spectroscopy study of lattice vibration and crystallographic orientation of monolayer MoS_2 under uniaxial strain. *Small* **2013**, *9*, 2857–2861. [CrossRef] [PubMed]
53. Molina-Sanchez, A.; Wirtz, L. Phonons in single-layer and few-layer MoS_2 and WS_2. *Phys. Rev. B* **2011**, *84*, 155413. [CrossRef]
54. Reale, F.; Palczynski, P.; Amit, I.; Jones, G.F.; Mehew, J.D.; Bacon, A.; Ni, N.; Sherrell, P.C.; Agnoli, S.; Craciun, M.F.; et al. High-mobility and high-optical quality atomically thin WS_2. *Sci. Rep.* **2017**, *7*, 14911. [CrossRef] [PubMed]
55. Gutierrez, H.R.; Perea-Lopez, N.; Elias, A.L.; Berkdemir, A.; Wang, B.; Lv, R.; Lopez-Urias, F.; Crespi, V.H.; Terrones, H.; Terrones, M. Extraordinary room-temperature photoluminescence in triangular WS_2 monolayers. *Nano Lett.* **2013**, *13*, 3447–3454. [CrossRef]
56. Liu, H.; Lu, J.; Ho, K.; Hu, Z.; Dang, Z.; Carvalho, A.; Tan, H.R.; Tok, E.S.; Sow, C.H. Fluorescence concentric triangles: A case of chemical heterogeneity in WS_2 atomic monolayer. *Nano Lett.* **2016**, *16*, 5559–5567. [CrossRef]
57. Feng, S.; Yang, R.; Jia, Z.; Xiang, J.; Wen, F.; Mu, C.; Nie, A.; Zhao, Z.; Xu, B.; Tao, C.; et al. Strain release induced novel fluorescence variation in CVD-grown monolayer WS_2 crystals. *ACS Appl. Mater. Interfaces* **2017**, *9*, 34071–34077. [CrossRef]
58. Liu, Z.; Amani, M.; Najmaei, S.; Xu, Q.; Zou, X.; Zhou, W.; Yu, T.; Qiu, C.; Birdwell, A.G.; Crowne, F.J.; et al. Strain and structure heterogeneity in MoS_2 atomic layers grown by chemical vapour deposition. *Nat. Commun.* **2014**, *5*, 5246. [CrossRef]

Article

Environmental Effects on the Electrical Characteristics of Back-Gated WSe_2 Field-Effect Transistors

Francesca Urban [1,2], Nadia Martucciello [2], Lisanne Peters [3], Niall McEvoy [3] and Antonio Di Bartolomeo [1,2,*]

1 Physics Department "E. R. Caianiello" and Interdepartmental Centre NanoMates, University of Salerno, via Giovanni Paolo II n. 132, 84084 Fisciano, Italy; furban@unisa.it
2 CNR-SPIN Salerno, via Giovanni Paolo II n. 132, 84084 Fisciano, Italy; nadia.martucciello@spin.cnr.it
3 AMBER & School of Chemistry, Trinity College Dublin, 2 Dublin, Ireland; petersli@tcd.ie (L.P.); nmcevoy@tcd.ie (N.M.E.)
* Correspondence: adibartolomeo@unisa.it; Tel.: +39-089-969189

Received: 4 October 2018; Accepted: 1 November 2018; Published: 3 November 2018

Abstract: We study the effect of polymer coating, pressure, temperature, and light on the electrical characteristics of monolayer WSe_2 back-gated transistors with Ni/Au contacts. Our investigation shows that the removal of a layer of poly(methyl methacrylate) (PMMA) or a decrease of the pressure change the device conductivity from p- to n-type. From the temperature behavior of the transistor transfer characteristics, a gate-tunable Schottky barrier at the contacts is demonstrated and a barrier height of ~70 meV in the flat-band condition is measured. We also report and discuss a temperature-driven change in the mobility and the subthreshold swing that is used to estimate the trap density at the WSe_2/SiO_2 interface. Finally, from studying the spectral photoresponse of the WSe_2, it is proven that the device can be used as a photodetector with a responsivity of $\sim 0.5\ AW^{-1}$ at 700 nm and 0.37 mW/cm^2 optical power.

Keywords: 2D materials; field effect transistors; PMMA; tungsten diselenide

1. Introduction

The continuous downscaling of the channel length and thickness in modern field effect transistors (FETs) has increased the need for atomically-layered materials to minimize short channel effects at extreme scaling limits [1,2]. Layered transition metal dichalcogenides (TMDs), owing to their two-dimensional structure, reasonable charge-carrier mobilities, and the absence of dangling bonds can enable extreme channel length scaling and have recently emerged as promising materials for future electronic and optoelectronic devices [3–7]. These graphene-like materials offer the advantages of sizeable and non-zero bandgap, high on/off ratio and quasi-ideal subthreshold swing, mechanical flexibility, and thermal and chemical stability. Similar to graphene, their electronic transport properties are strongly influenced by the choice of the metal contacts [8–10], by interface traps and impurities [11,12], as well as by structural defects and environmental exposure [13–17]. These effects need to be understood and controlled for technological applications.

Molybdenum disulfide (MoS_2) has been one of the most heavily investigated systems from the TMD family to date [18–25]. Similar to MoS_2, tungsten diselenide (WSe_2), whose electrical and optical properties have been relatively less explored [26], is characterized by an indirect bandgap (1.0–1.2 eV) in the bulk form and shows a transition to a direct gap of 1.6 eV when it is thinned to monolayer [27]. Recent reports on WSe_2 FETs have demonstrated a relatively high field-effect mobility controllable by temperature and bias voltage [26,28], an ideal subthreshold swing ~ 60 mV/dec [29] and an *on/off* ratio up to 10^8. The ambipolar behavior, controllable using different metal contacts, like In or Pd [8], which favor electron and hole injection, respectively [26,30,31], makes mono- and few-layer WSe_2

an interesting material for complementary logic applications; indeed, a stable WSe_2-based CMOS technology has been demonstrated [32–34].

A great challenge for electronic integration of WSe_2 is the development of low-resistance ohmic contacts, a task often complicated by the appearance of Schottky barriers due to the occurrence of Fermi level pinning [35,36]. Accordingly, several studies have aimed to clarify the role of the contacts, focusing on the carrier transport at the WSe_2/metal interface [30,37,38].

In this paper, we study back-gated monolayer WSe_2 devices with Ni contacts, measuring their electrical characteristics under different conditions, considering, for instance, the effect of a poly(methyl methacrylate) (PMMA) coating layer, and the dependence on the chamber pressure and the sample temperature. Similar to graphene [39,40], we observe that PMMA strongly influences the electrical transport, in this case to the extent that the polarity of the device changes from p-type to n-type conduction when the PMMA layer is removed. We demonstrate that lowering the pressure on air-exposed WSe_2 FETs affects their characteristics in a similar way to PMMA, turning the conduction from p- to n-type. Furthermore, from the current-voltage (I-V) characteristics measured at different temperatures, we prove a gate modulation of the Schottky barrier (SB) at the contacts.

In addition, we study the temperature dependence of the carrier mobility and the subthreshold swing and show that both undergo a change of behavior with increasing temperature. From the subthreshold swing data, we derive the interface trap density, which affects the photoresponse of the device. The monolayer WSe_2 device, characterized at several laser wavelengths, achieves a responsivity as high as $\sim 0.5\ AW^{-1}$ at 700 nm, i.e. at a photon energy close to the WSe_2 bandgap.

2. Experimental

The WSe_2 flakes were grown in a two-zone heating furnace. Selenium pellets (Sigma-Aldrich Inc, St. Louis, MO, USA) were evaporated at 250 °C in the lower-temperature, upstream heating zone, while in the high-temperature, downstream zone the tungsten precursor (20 nm sputtered and subsequently oxidized tungsten) was placed with the growth substrate. A highly p-doped Si (silicon) substrate covered by 300 nm of SiO_2 (silicon dioxide) was placed top down on the tungsten precursor, which was heated up to 850 °C. The tungsten precursor/growth substrate stack forms a microreactor, which increases the reactivity due to the close proximity between the precursor and growth substrate, requiring a lower amount of precursor and minimizing the contamination of the furnace. This is a similar approach to previous reports on the growth of MoS_2 in a microreactor [41] but in this case different chalcogen and transition metal precursors are used. Both furnaces were kept at the reaction temperatures for 40 min under a flow of 50 sccm forming gas (H_2/Ar 1:9) at a pressure of 6 Torr, after which the furnace was cooled down.

A schematic of the back-gated FET device and a scanning electron microscope top-view of a WSe_2 monolayer with evaporated Ni/Au (5/50 nm) contacts, made by use of e-beam lithography, are shown in Figure 1a,b. In the following, the transistor characterization refers to contact 1 and 2, which define a device with channel length L $\sim$ 2 μm and mean width W $\sim$ 22 μm (Figure 1b). The electrical analyses are performed using a Keithley 4200 SCS (semiconductor characterization system, Tektronix Inc., Beaverton, OR, USA) connected with a Janis ST-500 probe station (Janis Research Company LLC, Woburn, MA, USA), equipped with four probes used for the electrical connection to the drain and source Ni/Au terminals and to the Si back-gate of the device.

Raman and photoluminescence (PL) spectra were acquired using a WITec Alpha 300 tool (WITec GmbH, Ulm, Germany) with a 532 nm excitation laser. The Raman spectrum of the WSe_2, displayed in Figure 1c, exhibits two peaks around $\sim$ 250 cm^{-1} and $\sim$ 260 cm^{-1}, corresponding to an overlapping contribution from the in-plane vibrations of W and Se atoms ($E^1{}_{2g}$) and out-of-plane vibrations of Se atoms (A_{1g}), and to a second-order resonant Raman mode ($2\,LA\,(M)$) due to LA phonons at the M point in the Brillouin zone [42,43], respectively. The peak frequency positions are typical of a WSe_2 monolayer of thickness $d \sim 0.7$ nm [29]. The monolayer structure of the flake is further confirmed by the PL spectrum of Figure 1d, which shows an intense and narrow peak with maximum at $\sim$ 778 nm and FWHM of $\sim 21\ nm$. Such a peak corresponds to a bandgap of $\sim$ 1.59 eV, a value closer to that of a monolayer than to that of a bilayer. Hence, both Raman and PL spectra indicate that the flake is a monolayer.

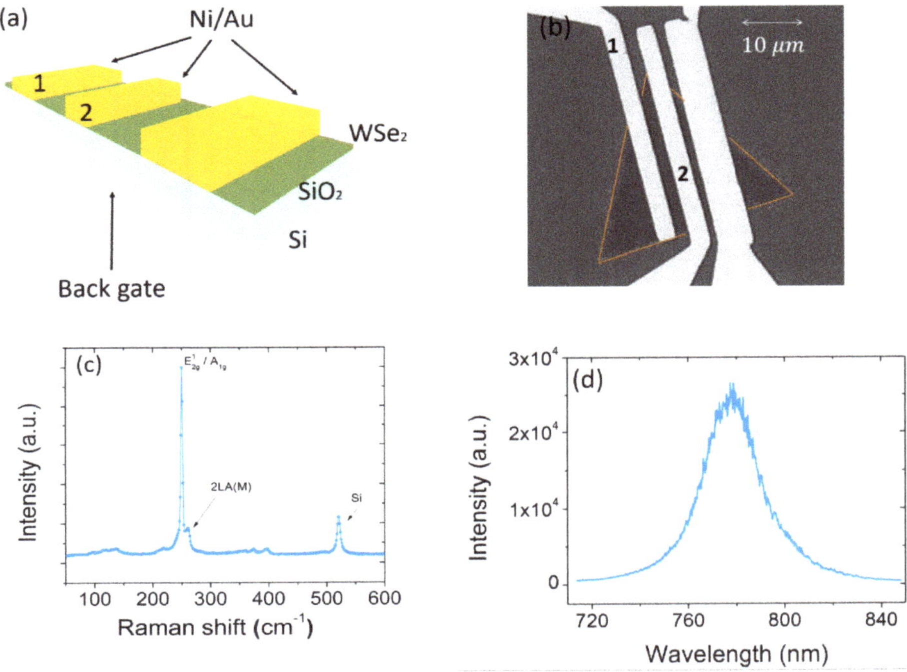

Figure 1. Schematic diagram (**a**) and optical microscope image (**b**) of the WSe_2 back gate FET transistor. Raman (**c**) and photoluminescence (**d**) spectra of the WSe_2 flake.

3. Results and Discussion

We start the transistor characterization by comparing the device I-V curves with and without a PMMA coating layer, which was used to protect the transistor channel from residue and adsorbates [44,45]. It has been observed that a PMMA film, or even only residue of it, can cause p-type doping of graphene and other 2D channels due to the presence of oxygen. Here, we report a similar effect for CVD-grown WSe_2 FETs, measured at $T = 293$ K, and $P \sim 2$ mbar.

The PMMA-covered devices behave like p-type transistors, as can be seen from the $I_{ds} - V_{gs}$ transfer curves of Figure 2a which show high channel current I_{ds} (*on*-state of the FET) at negative gate voltages, V_{gs}. The p-type conduction is explained considering the charge transfer to oxygen which, acting as electron capture center, suppresses the free electron density and enhances the hole concentration in the channel. Furthermore, the transition of the channel to p-type could cause a depinning of the Fermi level and facilitate hole injection at the contacts (indeed, in TMDs, Fermi level pinning often occurs close to the minimum of the conduction band) [36,46–48]. After the removal of the PMMA by immersion in acetone, a dramatic change to n-type behavior appeared, with the *on*-state at $V_{gs} < 0$ V, as shown in Figure 2b. A similar effect has been reported in literature [32,49] for exfoliated WSe_2 flakes on an SiO_2/Si substrate covered by F_4PCNQ-doped PMMA.

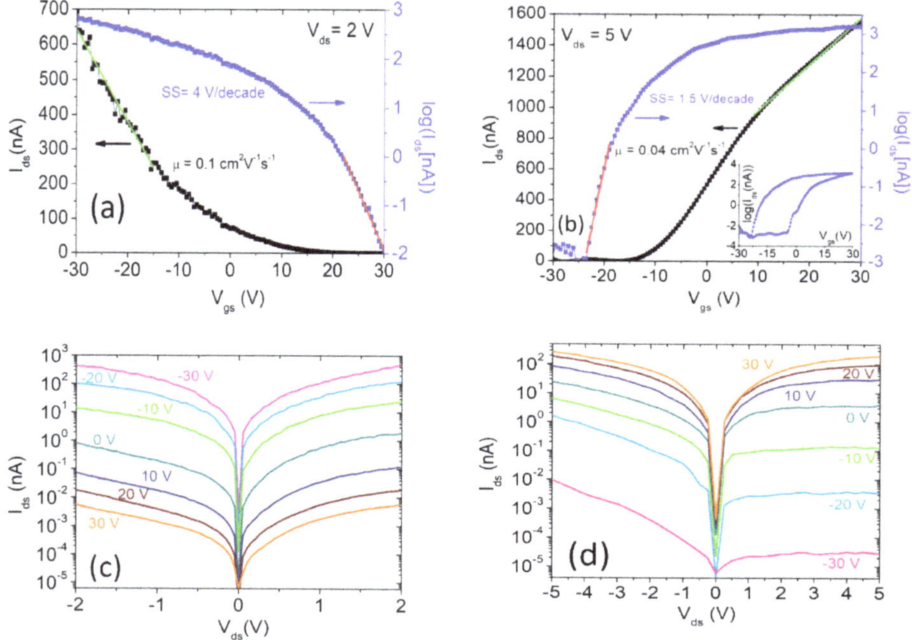

Figure 2. Transfer characteristics ($I_{ds} - V_{gs}$ curves) obtained at a drain voltage bias $V_{ds} = 2\ V$ for the device covered with PMMA (**a**) and after the removal of PMMA (**b**) at $V_{ds} = 5\ V$. The inset shows a complete cycle with the gate voltage V_{gs} swept forward and backward. Output characteristics ($I_{ds} - V_{ds}$ curves) at different gate voltages for the device with (**c**) and without (**d**) PMMA. For the uncovered device, the drain bias was increased from V_{ds} = 2 V to V_{ds} = 5 V to better characterize the above-threshold region.

The corresponding $I_{ds} - V_{ds}$ output characteristics are reported in Figure 2c,d. Both plots show non-linear behavior for the device in the *on* state with increasing positive-negative asymmetry when the device approaches the *off* state. This points to the presence of Schottky barriers at the Ni/WSe_2 contacts, possibly with slightly different heights [37,50].

For increasing V_{gs}, the channel current at constant V_{ds} shows an exponential dependence (below the threshold region) followed by a linear or power law behavior (above the threshold region).

A quadratic behavior is particularly evident in the transfer characteristic of Figure 2a, even though the transistor is operated in the triode region. The parabolic dependence of I_{ds} on V_{gs} can be ascribed to the linear gate-voltage dependence of the mobility μ [51,52], which defines the drain current as:

$$I_{ds} = \frac{WC_{ox}\,\mu}{L}(V_{gs} - V_{th})V_{ds} \tag{1}$$

with:

$$\mu = \mu_B(V_{gs} - V_{th}) \tag{2}$$

in which μ_B represents the mobility per unit gate voltage and V_{th} is the threshold voltage. The V_{gs}-dependent mobility can be explained by considering that the increasing carrier density becomes more effective at screening Coulomb scattering or in filling trap states at higher V_{gs}, thus resulting in enhanced mobility.

The dependence of the mobility on the gate voltage can be established by extracting it in the usual way using:

$$\mu = \frac{L}{W}\frac{1}{C_{ox}}\frac{1}{V_{ds}}\frac{dI_{ds}}{dV_{gs}} \tag{3}$$

Figure 3a,b show the $\mu - V_{gs}$ curves on logarithmic and linear (insets) scales obtained from Equation (3) and the data of Figure 2, for the devices with and without PMMA, respectively. These confirm a linear dependence of μ on V_{gs} over a certain range. Remarkably, for the device with removed PMMA, the mobility shows the typical decrease observed in common FETs due to increased scattering suffered by carriers attracted at the channel/dielectric interface at higher gate voltages.

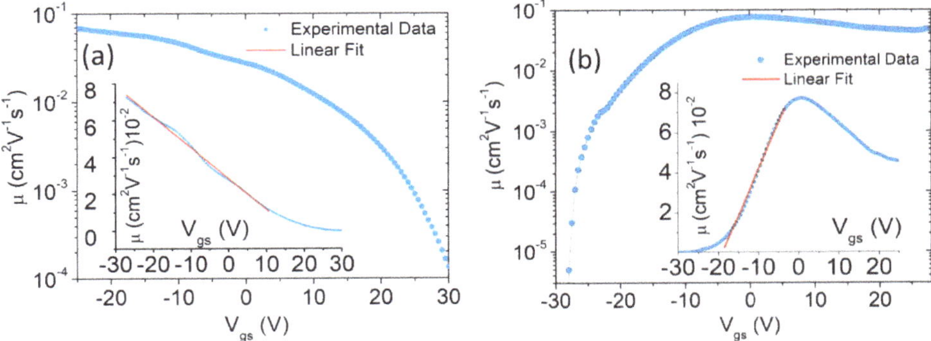

Figure 3. Mobility versus gate voltage on a logarithmic scale for the WSe_2 flake covered (**a**) and uncovered (**b**) by PMMA. The inset graphs show the mobility on a linear scale.

By neglecting the V_{gs} dependence of the mobility, as is usually done in the literature, μ can be obtained by fitting a straight line to the transfer characteristics, as shown in Figure 2a,b.

By this method, we estimate an electron mobility of $\sim 0.04\ cm^2V^{-1}s^{-1}$ for the n-type transistor without PMMA, consistent with other works with WSe_2 on SiO_2 [41], and a hole mobility of $\sim 0.1\ \mathrm{cm^2V^{-1}s^{-1}}$ for the PMMA-covered p-type transistor. We notice that, although a different channel carrier concentration might contribute to this difference, these values are consistent with the higher hole mobility in WSe_2 reported elsewhere [29,32,45]. The low mobility indicates a high density of trap states, which are also responsible for the hysteretic behavior of the transfer characteristic shown in the inset of Figure 2b. The hysteresis is caused by trapping and detrapping of charge carriers, whose potential adds to that of the back-gate [51–54].

The subthreshold swing, $SS = dV_{gs}/dlog(I_{ds})$, is 4 $V/decade$ and 1.5 $V/decade$, for the p-type and n-type transistor, respectively. The different SS results from a different trap density at the WSe_2/dielectric interface, implying a higher trap density when the WSe_2 channel is covered by PMMA, which acts as a second interface [55,56].

After the removal of the polymeric film and exposure of the device to air for a few days, we observed a restoration of a prevailing p-type behavior due to O_2 and water adsorption on the WSe_2 surface and possible depinning of the Fermi level. We then studied the effect of dynamic pressure by increasing the vacuum level of the probe station chamber from atmosphere ($\sim 1\ bar$) to $\sim 10^{-5}$ mbar. As reported in Figure 4a, the transistor transfer characteristic changed again from p- to n-type with a gradual decrease of the subthreshold swing and an increase of the *on/off* ratio, as shown in Figure 4b. We attribute the polarity change to the desorption of adsorbed O_2 and H_2O and to the consequent possible pinning of the Fermi level close to the minimum of the conduction band.

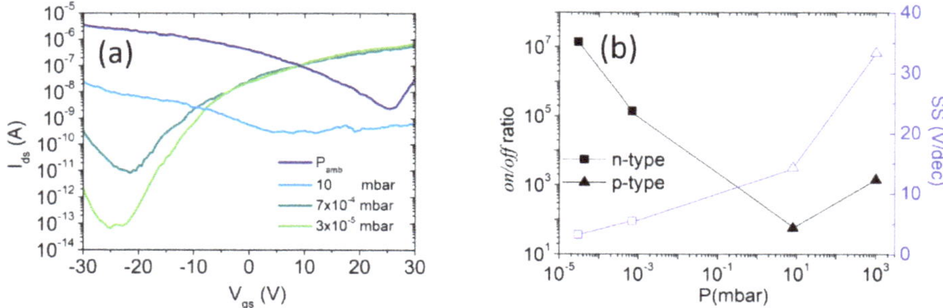

Figure 4. (**a**) Transfer characteristics at different pressures from atmospheric value (blue curve) to $\sim 10^{-5}$ mbar (light green curve) (**b**) *On/Off* ratio (full marks, left scale) and subthreshold swing (empty marks, right scale) as a function of the chamber pressure.

We then examined the temperature (T) dependence of the transfer characteristics of the PMMA-free, n-type transistor at low pressure, which can be conveniently used to investigate the Schottky barrier for electrons at the contacts.

We extract the Schottky barrier at the flat-band condition from a plot of the Schottky barrier height as a function of V_{gs} for the device at a source-drain bias of 5 V (Figure 5). Given that the device is n-type, such a barrier refers to electron injection from the contacts and it is caused by the aforementioned pinning of the Fermi level close to the minimum of the conduction band. Measuring the $I_{ds} - V_{gs}$ characteristics of the device at several temperatures (Figure 5a) and extracting $I_{ds} - T$ datasets at given gate voltages (examples are marked by the vertical lines in Figure 5a), we constructed the Arrhenius plot of Figure 5b, showing the $ln\left(I_{ds}/T^{3/2}\right) - \frac{1}{T}$ curves at a representative subset of V_{gs} values. We assumed that the contacts behave as two back-to-back Schottky junctions, where the current is controlled by the reverse-biased junction and is written as:

$$I_{ds} \sim T^{3/2}\, exp\left(-\frac{\Phi_B}{kT}\right) \tag{4}$$

where k is the Boltzmann constant, T is the absolute temperature, and Φ_B is the Schottky barrier height [30,57,58]. According to equation (4), a linear fit of $\ln\left(I_{ds}/T^{3/2}\right)$ vs $1/T$ for each V_{gs} dataset in Figure 5b yields a Schottky barrier Φ_B. The so-obtained $\Phi_B - V_{gs}$ relationship is displayed in Figure 5c and can be divided into three zones, each one corresponding to a different transport regime, consistent with the behavior of the transfer characteristics of Figure 5a.

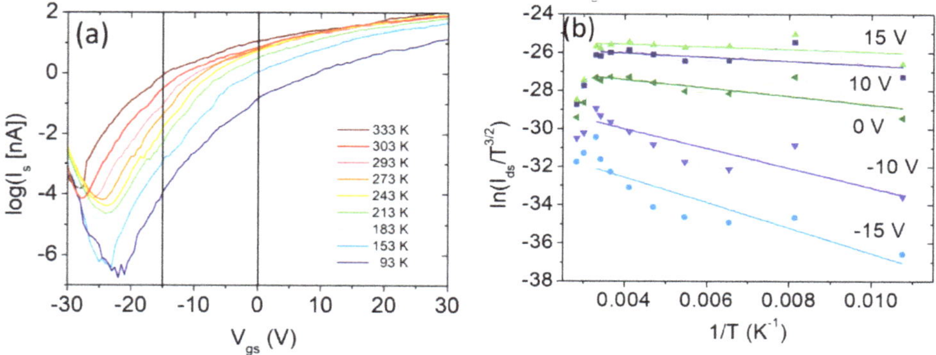

Figure 5. *Cont.*

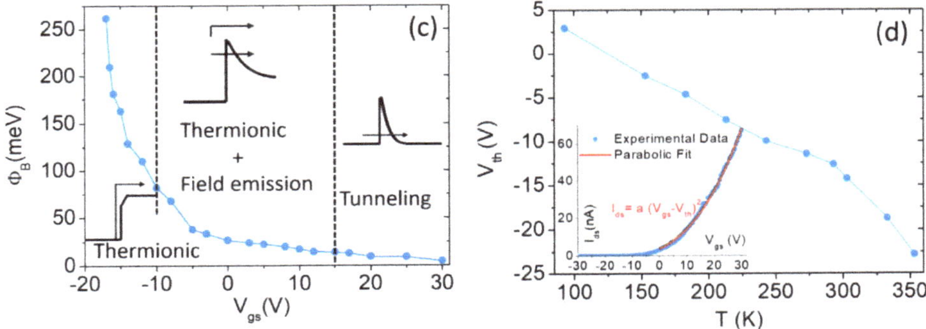

Figure 5. (**a**) Transfer characteristic at several temperatures. (**b**) Arrhenius plot of the current at different temperatures corresponding to a subset of the gate voltages (two of these V_{gs} values are represented by the vertical lines in (**a**)). (**c**) Apparent Schottky barrier as a function of the gate voltage; the insets show the band alignment and the transport regimes at the Ni/WSe_2 reverse-biased contact. (**d**) Threshold voltage V_{th} as a function of the temperature; the inset shows, as an example, the parabolic fit of the $I_{ds} - V_{gs}$ curve at $T = 273\ K$.

At low gate voltage the device is set in the *off* state and the transport is due to the thermal excitation of electrons over the barrier. The WSe_2 conduction-band level is gradually lowered by the increasing gate voltage, as sketched in the insets of Figure 5c. This results in a lowering of the barrier with a subsequent steep exponential rise of the current in the transfer characteristic (with 60 $mV/decade$ slope in the ideal case). When the gate voltage is further increased the device reaches the flat band condition ($V_{gs} = V_{FB}$), which sometimes appears in the subthreshold part of the transfer characteristics as a sudden change of slope; for $V_{gs} > V_{FB}$ the device enters the so-called Schottky regime which includes part of the downward bended region of $I_{ds} - V_{ds}$ curves and is characterized by thermionic emission and field emission. Finally, at higher V_{gs}, tunneling through the thinned Ni/WSe_2 barrier becomes the dominant conduction mechanism and the device reaches the above threshold region with a linear, or power-law, $I_{ds} - V_{gs}$ dependence.

The gate voltage that corresponds to V_{FB} is identified by the change of slope in the $\Phi_B - V_{gs}$ plot at lower V_{gs}. The Φ_B corresponding to $V_{gs} = V_{FB}$ is the so-called Schottky barrier height at the flat-band (or simply Schottky barrier). From the plot in Figure 5c, its value is $\sim 70\ meV$, confirming the presence of a barrier at the Ni-WSe_2 contacts, inferred from the asymmetric output characteristics of Figure 2.

Figure 5d shows the temperature-dependent behavior of the threshold voltage V_{th}, which has been extracted assuming a quadratic $I_{ds} - V_{gs}$ law as expressed by Equation (1) and (2). The decrease in V_{th} is easily explained by considering that the increasing temperature accelerates the transition from the Schottky to the power-law (above threshold) regime; furthermore, the plot seems to indicate a change of slope above room temperature.

Figure 6a reports the temperature-dependent behavior of the mobility, μ, at $V_{gs} = 10\ V$ obtained from the quadratic fit of the $I_{ds} - V_{gs}$ curves.

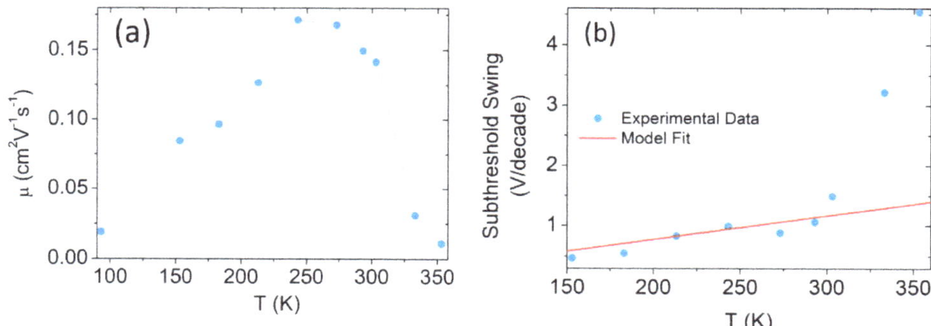

Figure 6. Temperature dependence of (**a**) mobility per unit voltage μ_B and (**b**) subthreshold swing.

The mobility increases for $T < 250$ K and decreases for $T > 250$ K, behavior typical of semiconductor materials, indicating that charged-impurity Coulomb scattering dominates at lower temperatures, while phonon scattering becomes the conduction-limiting mechanism at high temperature [44].

The subthreshold swing has a dependence on temperature that can be simplified with the following expression:

$$SS = n\frac{kT}{q}ln10 \tag{5}$$

where n is the body factor which is related to the interface trap (C_{it}), SiO_2 (C_{SiO2}) and channel depletion layer (C_{dl}) capacitances by:

$$n = 1 + \frac{C_{it} + C_{dl}}{C_{ox}} \tag{6}$$

Figure 6b confirms the linear $SS - T$ dependence (Equation 5) but shows an unexpected rise above room temperature. The deviation from Equation (5) behavior at high temperature is a consequence of the low Schottky barrier which becomes less effective above room temperature ($kT = 26$ meV), resulting in an increase of the subthreshold current leakage.

Assuming that the WSe_2 monolayer channel is fully depleted, i.e. that $C_{dl} \approx 0$, from the fit of the experimental data with Equation (5), we obtain a $n \approx 48$ and an interface trap density $N_{it} = \frac{C_{it}}{q^2} \approx 1.3 \times 10^{13}\ eV^{-1}cm^{-2}$, which is consistent with previous results reported in the literature [59].

The presence of such a density of trap states explains the observed hysteretic behavior of the transfer characteristic, displayed in the inset of Figure 2b [52]. It also affects the electrical response of the device under illumination.

We performed photocurrent measurements with light at different wavelengths, selected by filtering a supercontinuous laser source (NKT Photonics, Superk Compact, wavelength ranging from 450 nm to 2400 nm, total output power of 110 mW) using pass-band filters with 50 nm bandwidth. Figure 7a shows the photoresponse of the WSe_2 FET to laser light pulses of 30 s for five different wavelengths.

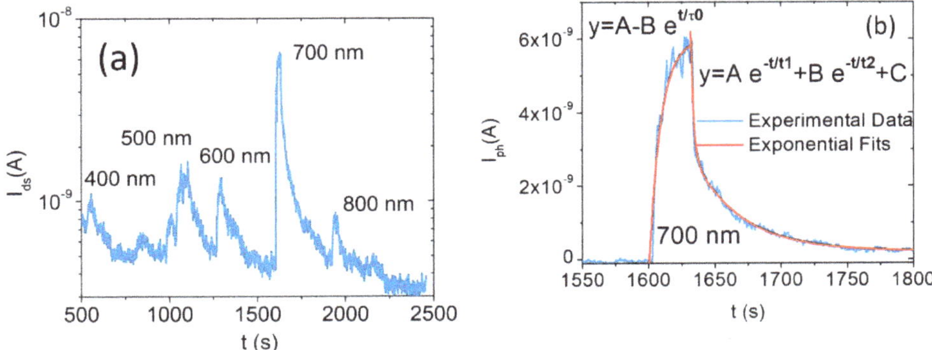

Figure 7. (**a**) Drain-source current measured under 30 s laser pulses at different wavelengths ($V_{ds} = +5\ V, V_{gs} = 0\ V,\ P \sim 10^{-4}\ mbar$). (**b**) Photocurrent generated by a 30 s laser pulse at the wavelength of $\sim 700\ nm$ and optical power $\sim 0.37\ mW/cm^2$ with exponential fits.

The photocurrent exhibits a higher peak at the wavelength of 700 nm (photon energy 1.7 eV), which is slightly above the bandgap of a WSe_2 monolayer, supporting the Raman and the PL spectroscopy assignment of the single-layer nature of the WSe_2 channel.

Figure 7b reports the photocurrent, $I_{ph} = I_{light} - I_{dark}$, obtained in response to a laser pulse of 30 s at a wavelength of $\sim 700\ nm$, and an optical power $\sim 0.37\ mW/cm^2$. It corresponds to a peak with rising time $\tau_0\ \sim 9\ s$ and a double exponential decay with times $\tau_1\ \sim 2\ s$ and $\tau_2\ \sim 36\ s$, indicating the presence of faster and slower traps [60]. Such features are consistent with a photoresponse decay longer than 5 s for quasi-ohmic contacts measured on similar WSe_2 FETs [61,62]. Indeed, we notice that the contact type can play an important role in the response time of WSe_2 phototransistors and that reduced times have been reported for Schottky contacts [61,62].

Furthermore, we estimate a photoresponsivity (Figure 7b):

$$R = \frac{I_{ph}}{W_{opt}} \approx 0.5\frac{A}{W} \tag{7}$$

where W_{opt} is the incident power. This is in good agreement with the previously reported value of 0.6 A/W obtained at 750 nm [62]. Such a responsivity is competitive with solid state devices on the market and, despite the ultrathin nature of the absorber, confirms the excellent photoresponse of monolayer WSe_2 due to its direct bandgap [63,64].

4. Conclusions

We showed that different environmental conditions can have dramatic effects on the electrical properties of back-gated transistors with monolayer WSe_2 channels. In particular, we demonstrated that the removal of a polymer coating layer, as well as of oxygen and water adsorbates, can change the conduction from p- to n-type. From I-V characterization at different temperatures, we extracted the Ni/WSe_2 Schottky barrier height, which we studied as a function of the back-gate voltage. We reported and discussed a change in the temperature behavior of the mobility and the subthreshold swing. Finally, we studied the photoresponse of the device to selected laser wavelengths achieving a responsivity competitive with solid-state devices on the market.

Author Contributions: F.U. and A.D.B. wrote the paper; A.D.B. and F.U. conceived and designed the experiments; F.U. and L.P. performed the experiments; F.U., N.M.E., and A.D.B. analyzed the data; and N.M., A.D.B., and N.M.E. contributed reagents/materials/analysis tools.

Funding: This research was funded by Regione Campania grant "POR Campania FSE 2014–2020, Asse III Ob. specifico l4, D.D. n. 80 del 31/05/2016", by CNR-SPIN grant "SEED Project 2017" and by Science Foundation Ireland grant number: 15/SIRG/3329.

Conflicts of Interest: The authors declare no conflict of interest.

References

1. Georgiou, T.; Jalil, R.; Belle, B.D.; Britnell, L.; Gorbachev, R.V.; Morozov, S.V.; Kim, Y.-J.; Gholinia, A.; Haigh, S.J.; Makarovsky, O.; et al. Vertical field-effect transistor based on graphene–WS_2 heterostructures for flexible and transparent electronics. *Nat. Nanotechnol.* **2012**, *8*, 100–103. [CrossRef] [PubMed]
2. Liu, H.; Neal, A.T.; Ye, P.D. Channel length scaling of MoS_2 MOSFETs. *ACS Nano* **2012**, *6*, 8563–8569. [CrossRef] [PubMed]
3. Lopez-Sanchez, O.; Lembke, D.; Kayci, M.; Radenovic, A.; Kis, A. Ultrasensitive photodetectors based on monolayer MoS_2. *Nat. Nanotechnol.* **2013**, *8*, 497. [CrossRef] [PubMed]
4. Pospischil, A.; Furchi, M.M.; Mueller, T. Solar-energy conversion and light emission in an atomic monolayer p–n diode. *Nat. Nanotechnol.* **2014**, *9*, 257. [CrossRef] [PubMed]
5. Radisavljevic, B.; Radenovic, A.; Brivio, J.; Giacometti, V.; Kis, A. Single-layer MoS_2 transistors. *Nat. Nanotechnol.* **2011**, *6*, 147. [CrossRef] [PubMed]
6. Baugher, B.W.H.; Churchill, H.O.H.; Yang, Y.; Jarillo-Herrero, P. Optoelectronic devices based on electrically tunable p–n diodes in a monolayer dichalcogenide. *Nat. Nanotechnol.* **2014**, *9*, 262. [CrossRef] [PubMed]
7. Ross, J.S.; Klement, P.; Jones, A.M.; Ghimire, N.J.; Yan, J.; Mandrus, D.G.; Taniguchi, T.; Watanabe, K.; Kitamura, K.; Yao, W.; et al. Electrically tunable excitonic light-emitting diodes based on monolayer WSe_2 p–n junctions. *Nat. Nanotechnol.* **2014**, *9*, 268. [CrossRef] [PubMed]
8. Liu, W.; Kang, J.; Sarkar, D.; Khatami, Y.; Jena, D.; Banerjee, K. Role of metal contacts in designing high-performance monolayer n-type WSe_2 field effect transistors. *Nano Lett.* **2013**, *13*, 1983–1990. [CrossRef] [PubMed]
9. Guo, Y.; Han, Y.; Li, J.; Xiang, A.; Wei, X.; Gao, S.; Chen, Q. Study on the resistance distribution at the contact between molybdenum disulfide and metals. *ACS Nano* **2014**, *8*, 7771–7779. [CrossRef] [PubMed]
10. Rai, A.; Movva, C.H.; Roy, A.; Taneja, D.; Chowdhury, S.; Banerjee, K.S. Progress in contact, doping and mobility engineering of MoS_2: An atomically thin 2D semiconductor. *Crystals* **2018**, *8*, 316. [CrossRef]
11. Dean, C.R.; Young, A.F.; Meric, I.; Lee, C.; Wang, L.; Sorgenfrei, S.; Watanabe, K.; Taniguchi, T.; Kim, P.; Shepard, K.L.; et al. Boron nitride substrates for high-quality graphene electronics. *Nat. Nanotechnol.* **2010**, *5*, 722. [CrossRef] [PubMed]
12. Chen, J.-H.; Jang, C.; Adam, S.; Fuhrer, M.S.; Williams, E.D.; Ishigami, M. Charged-impurity scattering in graphene. *Nat. Phys.* **2008**, *4*, 377. [CrossRef]
13. Lin, J.; Zhong, J.; Zhong, S.; Li, H.; Zhang, H.; Chen, W. Modulating electronic transport properties of MoS_2 field effect transistor by surface overlayers. *Appl. Phys. Lett.* **2013**, *103*, 063109. [CrossRef]
14. Mouri, S.; Miyauchi, Y.; Matsuda, K. Tunable photoluminescence of monolayer MoS_2 via chemical doping. *Nano Lett.* **2013**, *13*, 5944–5948. [CrossRef] [PubMed]
15. Late, D.J.; Liu, B.; Matte, H.S.S.R.; Dravid, V.P.; Rao, C.N.R. Hysteresis in single-layer MoS_2 field effect transistors. *ACS Nano* **2012**, *6*, 5635–5641. [CrossRef] [PubMed]
16. Roy, A.; Ghosh, R.; Rai, A.; Sanne, A.; Kim, K.; Movva, H.C.P.; Dey, R.; Pramanik, T.; Chowdhury, S.; Tutuc, E.; et al. Intra-domain periodic defects in monolayer MoS_2. *Appl. Phys. Lett.* **2017**, *110*, 201905. [CrossRef]
17. Zhao, P.; Kiriya, D.; Azcatl, A.; Zhang, C.; Tosun, M.; Liu, Y.-S.; Hettick, M.; Kang, J.S.; McDonnell, S.; KC, S.; et al. Air stable p-doping of WSe_2 by covalent functionalization. *ACS Nano* **2014**, *8*, 10808–10814. [CrossRef] [PubMed]
18. Urban, F.; Passacantando, M.; Giubileo, F.; Iemmo, L.; Di Bartolomeo, A. Transport and field emission properties of MoS_2 bilayers. *Nanomaterials* **2018**, *8*, 151. [CrossRef] [PubMed]
19. Fang, H.; Tosun, M.; Seol, G.; Chang, T.C.; Takei, K.; Guo, J.; Javey, A. Degenerate n-doping of few-layer transition metal dichalcogenides by potassium. *Nano Lett.* **2013**, *13*, 1991–1995. [CrossRef] [PubMed]

20. Rai, A.; Valsaraj, A.; Movva, H.C.P.; Roy, A.; Ghosh, R.; Sonde, S.; Kang, S.; Chang, J.; Trivedi, T.; Dey, R.; et al. Air stable doping and intrinsic mobility enhancement in monolayer molybdenum disulfide by amorphous titanium suboxide encapsulation. *Nano Lett.* **2015**, *15*, 4329–4336. [CrossRef] [PubMed]
21. Tongay, S.; Zhou, J.; Ataca, C.; Lo, K.; Matthews, T.S.; Li, J.; Grossman, J.C.; Wu, J. Thermally driven crossover from indirect toward direct bandgap in 2D semiconductors: $MoSe_2$ versus MoS_2. *Nano Lett.* **2012**, *12*, 5576–5580. [CrossRef] [PubMed]
22. Jariwala, D.; Sangwan, V.K.; Late, D.J.; Johns, J.E.; Dravid, V.P.; Marks, T.J.; Lauhon, L.J.; Hersam, M.C. Band-like transport in high mobility unencapsulated single-layer MoS_2 transistors. *Appl. Phys. Lett.* **2013**, *102*, 173107. [CrossRef]
23. Valsaraj, R.A.; Movva, H.C.P.; Roy, A.; Tutuc, E.; Register, L.F.; Banerjee, S.K. Interfacial-oxygen-vacancy mediated doping of MoS2 by high-κ dielectrics. In *2015 73rd Annual Device Research Conference (DRC), Proceedings of the 73rd Annual Device Research Conference (DRC), Columbus, OH, USA, 21–24 June 2015*; IEEE: Piscataway, NJ, USA, 06 August 2015; pp. 189–190.
24. Desai, S.B.; Madhvapathy, S.R.; Sachid, A.B.; Llinas, J.P.; Wang, Q.; Ahn, G.H.; Pitner, G.; Kim, M.J.; Bokor, J.; Hu, C.; et al. MoS_2 transistors with 1-nanometer gate lengths. *Science* **2016**, *354*, 99. [CrossRef] [PubMed]
25. Yang, L.M.; Majumdar, K.; Du, Y.; Liu, H.; Wu, H.; Hatzistergos, M.; Hung, P.Y.; Tieckelmann, R.; Tsai, W.; Hobbs, C.; et al. High-performance MoS_2 field-effect transistors enabled by chloride doping: Record low contact resistance (0.5 kΩ·μm) and record high drain current (460 μA/μm). In *2014 Symposium on VLSI Technology (VLSI-Technology): Digest of Technical Papers, Proceedings of 2014 Symposium on VLSI Technology, Honolulu, HI, USA, 9–12 June 2014*; IEEE: Piscataway, NJ, USA, 11 September 2014; pp. 1–2.
26. Allain, A.; Kis, A. Electron and hole mobilities in single-layer WSe_2. *ACS Nano* **2014**, *8*, 7180–7185. [CrossRef]
27. Zhao, W.; Ghorannevis, Z.; Chu, L.; Toh, M.; Kloc, C.; Tan, P.-H.; Eda, G. Evolution of electronic structure in atomically thin sheets of WS_2 and WSe_2. *ACS Nano* **2013**, *7*, 791–797. [CrossRef]
28. Pradhan, N.R.; Rhodes, D.; Memaran, S.; Poumirol, J.M.; Smirnov, D.; Talapatra, S.; Feng, S.; Perea-Lopez, N.; Elias, A.L.; Terrones, M.; et al. Hall and field-effect mobilities in few layered p-WSe_2 field-effect transistors. *Sci. Rep.* **2015**, *5*, 8979. [CrossRef] [PubMed]
29. Fang, H.; Chuang, S.; Chang, T.C.; Takei, K.; Takahashi, T.; Javey, A. High-performance single layered WSe_2 p-FETs with chemically doped contacts. *Nano Lett.* **2012**, *12*, 3788–3792. [CrossRef] [PubMed]
30. Das, S.; Appenzeller, J. WSe_2 field effect transistors with enhanced ambipolar characteristics. *Appl. Phys. Lett.* **2013**, *103*, 103501. [CrossRef]
31. Liu, B.; Fathi, M.; Chen, L.; Abbas, A.; Ma, Y.; Zhou, C. Chemical vapor deposition growth of monolayer WSe_2 with tunable device characteristics and growth mechanism study. *ACS Nano* **2015**, *9*, 6119–6127. [CrossRef]
32. Yu, L.; Zubair, A.; Santos, E.J.G.; Zhang, X.; Lin, Y.; Zhang, Y.; Palacios, T. High-performance WSe_2 complementary metal oxide semiconductor technology and integrated circuits. *Nano Lett.* **2015**, *15*, 4928–4934. [CrossRef]
33. Tosun, M.; Chuang, S.; Fang, H.; Sachid, A.B.; Hettick, M.; Lin, Y.; Zeng, Y.; Javey, A. High-gain inverters based on WSe_2 complementary field-effect transistors. *ACS Nano* **2014**, *8*, 4948–4953. [CrossRef] [PubMed]
34. Das, S.; Dubey, M.; Roelofs, A. High gain, low noise, fully complementary logic inverter based on bi-layer WSe_2 field effect transistors. *Appl. Phys. Lett.* **2014**, *105*, 083511. [CrossRef]
35. Giubileo, F.; Di Bartolomeo, A. The role of contact resistance in graphene field-effect devices. *Prog. Surf. Sci.* **2017**, *92*, 143–175. [CrossRef]
36. Di Bartolomeo, A.; Giubileo, F.; Romeo, F.; Sabatino, P.; Carapella, G.; Iemmo, L.; Schroeder, T.; Lupina, G. Graphene field effect transistors with niobium contacts and asymmetric transfer characteristics. *Nanotechnology* **2015**, *26*, 475202. [CrossRef] [PubMed]
37. Di Bartolomeo, A.; Grillo, A.; Urban, F.; Iemmo, L.; Giubileo, F.; Luongo, G.; Amato, G.; Croin, L.; Sun, L.; Liang, S.J.; et al. Asymmetric schottky contacts in bilayer MoS_2 field effect transistors. *Adv. Funct. Mater.* **2018**, *28*, 1800657. [CrossRef]
38. Chuang, H.-J.; Tan, X.; Ghimire, N.J.; Perera, M.M.; Chamlagain, B.; Cheng, M.M.-C.; Yan, J.; Mandrus, D.; Tománek, D.; Zhou, Z. High mobility WSe_2 p- and n-type field-effect transistors contacted by highly doped graphene for low-resistance contacts. *Nano Lett.* **2014**, *14*, 3594–3601. [CrossRef] [PubMed]

39. Borin Barin, G.; Song, Y.; de Fátima Gimenez, I.; Souza Filho, A.G.; Barreto, L.S.; Kong, J. Optimized graphene transfer: Influence of polymethylmethacrylate (PMMA) layer concentration and baking time on graphene final performance. *Carbon* **2015**, *84*, 82–90. [CrossRef]
40. Son, B.H.; Kim, H.S.; Jeong, H.; Park, J.-Y.; Lee, S.; Ahn, Y.H. Electron beam induced removal of PMMA layer used for graphene transfer. *Sci. Rep.* **2017**, *7*, 18058. [CrossRef] [PubMed]
41. O'Brien, M.; McEvoy, N.; Hallam, T.; Kim, H.-Y.; Berner, N.C.; Hanlon, D.; Lee, K.; Coleman, J.N.; Duesberg, G.S. Transition metal dichalcogenide growth via close proximity precursor supply. *Sci. Rep.* **2014**, *4*, 7374. [CrossRef] [PubMed]
42. Zhao, W.; Ghorannevis, Z.; Amara, K.K.; Pang, J.R.; Toh, M.; Zhang, X.; Kloc, C.; Tan, P.H.; Eda, G. Lattice dynamics in mono- and few-layer sheets of WS_2 and WSe_2. *Nanoscale* **2013**, *5*, 9677–9683. [CrossRef] [PubMed]
43. O'Brien, M.; McEvoy, N.; Hanlon, D.; Hallam, T.; Coleman, J.N.; Duesberg, G.S. Mapping of low-frequency raman modes in CVD-grown transition metal dichalcogenides: Layer number, stacking orientation and resonant effects. *Sci. Rep.* **2016**, *6*, 19476. [CrossRef] [PubMed]
44. Ovchinnikov, D.; Allain, A.; Huang, Y.-S.; Dumcenco, D.; Kis, A. Electrical transport properties of single-layer WS_2. *ACS Nano* **2014**, *8*, 8174–8181. [CrossRef] [PubMed]
45. Schmidt, H.; Giustiniano, F.; Eda, G. Electronic transport properties of transition metal dichalcogenide field-effect devices: Surface and interface effects. *Chem. Soc. Rev.* **2015**, *44*, 7715–7736. [CrossRef] [PubMed]
46. Pandey, A.; Prasad, A.; Moscatello, J.; Ulmen, B.; Yap, Y.K. Enhanced field emission stability and density produced by conical bundles of catalyst-free carbon nanotubes. *Carbon* **2010**, *48*. [CrossRef]
47. Di Bartolomeo, A.; Giubileo, F.; Iemmo, L.; Romeo, F.; Russo, S.; Unal, S.; Passacantando, M.; Grossi, V.; Cucolo, A.M. Leakage and field emission in side-gate graphene field effect transistors. *Appl. Phys. Lett.* **2016**, *109*, 023510. [CrossRef]
48. Pirkle, A.; Chan, J.; Venugopal, A.; Hinojos, D.; Magnuson, C.W.; McDonnell, S.; Colombo, L.; Vogel, E.M.; Ruoff, R.S.; Wallace, R.M. The effect of chemical residues on the physical and electrical properties of chemical vapor deposited graphene transferred to SiO_2. *Appl. Phys. Lett.* **2011**, *99*, 122108. [CrossRef]
49. Movva, H.C.P.; Rai, A.; Kang, S.; Kim, K.; Fallahazad, B.; Taniguchi, T.; Watanabe, K.; Tutuc, E.; Banerjee, S.K. High-mobility holes in dual-gated WSe_2 field-effect transistors. *ACS Nano* **2015**, *9*, 10402–10410. [CrossRef] [PubMed]
50. Di Bartolomeo, A.; Urban, F.; Passacantando, M.; McEvoy, N.; Peters, L.; Iemmo, L.; Luongo, G.; Romeo, F.; Giubileo, F. A WSe_2 vertical field emission transistor. *arXiv*, 2018; arXiv:1808.02127.
51. Gupta, D.; Katiyar, M.; Gupta, D. Mobility estimation incorporating the effects of contact resistance and gate voltage dependent mobility in top contact organic thin film transistors. In Proceedings of the ASID 2006, New Delhi, India, 8–12 October 2006; pp. 425–428.
52. Di Bartolomeo, A.; Genovese, L.; Giubileo, F.; Iemmo, L.; Luongo, G.; Tobias, F.; Schleberger, M. Hysteresis in the transfer characteristics of MoS_2 transistors. *2D Materials* **2018**, *5*, 015014. [CrossRef]
53. Guo, Y.; Wei, X.; Shu, J.; Liu, B.; Yin, J.; Guan, C.; Han, Y.; Gao, S.; Chen, Q. Charge trapping at the MoS_2-SiO_2 interface and its effects on the characteristics of MoS_2 metal-oxide-semiconductor field effect transistors. *Appl. Phys. Lett.* **2015**, *106*, 103109. [CrossRef]
54. Di Bartolomeo, A.; Yang, Y.; Rinzan, M.B.M.; Boyd, A.K.; Barbara, P. Record endurance for single-walled carbon nanotube–based memory cell. *Nanoscale Res. Lett.* **2010**, *5*, 1852. [CrossRef] [PubMed]
55. Gatensby, R.; Hallam, T.; Lee, K.; McEvoy, N.; Duesberg, G.S. Investigations of vapour-phase deposited transition metal dichalcogenide films for future electronic applications. *Solid State Electron.* **2016**, *125*, 39–51. [CrossRef]
56. Giubileo, F.; Iemmo, L.; Passacantando, M.; Urban, F.; Luongo, G.; Sun, L.; Amato, G.; Enrico, E.; Di Bartolomeo, A. Effect of electron irradiation on the transport and field emission properties of few-layer MoS_2 field effect transistors. *arXiv*, 2018; arXiv:1808.01185.
57. Das, S.; Chen, H.-Y.; Penumatcha, A.V.; Appenzeller, J. High performance multilayer MoS_2 transistors with scandium contacts. *Nano Lett.* **2013**, *13*, 100–105. [CrossRef] [PubMed]
58. Houssa, M.; Dimoulas, A.; Molle, A. *2D Materials for Nanoelectronics*. 2016. Available online: https://inn.demokritos.gr/wp-content/uploads/2016/05/Promo_K24702_Flyer.pdf (accessed on 1 October 2018).

59. Wang, J.; Rhodes, D.; Feng, S.; Nguyen, M.A.T.; Watanabe, K.; Taniguchi, T.; Mallouk, T.E.; Terrones, M.; Balicas, L.; Zhu, J. Gate-modulated conductance of few-layer WSe_2 field-effect transistors in the subgap regime: Schottky barrier transistor and subgap impurity states. *Appl. Phys. Lett.* **2015**, *106*, 152104. [CrossRef]
60. Di Bartolomeo, A.; Genovese, L.; Foller, T.; Giubileo, F.; Luongo, G.; Croin, L.; Liang, S.-J.; Ang, L.K.; Schleberger, M. Electrical transport and persistent photoconductivity in monolayer MoS_2 phototransistors. *Nanotechnology* **2017**, *28*, 214002. [CrossRef] [PubMed]
61. Zhang, W.; Chiu, M.-H.; Chen, C.-H.; Chen, W.; Li, L.-J.; Wee, A.T.S. Role of metal contacts in high-performance phototransistors based on WSe_2 monolayers. *ACS Nano* **2014**, *8*, 8653–8661. [CrossRef] [PubMed]
62. Wang, T.; Andrews, K.; Bowman, A.; Hong, T.; Koehler, M.; Yan, J.; Mandrus, D.; Zhou, Z.; Xu, Y.-Q. High-performance WSe_2 phototransistors with 2D/2D ohmic contacts. *Nano Lett.* **2018**, *18*, 2766–2771. [CrossRef] [PubMed]
63. Li, H.; Qin, M.; Wang, L.; Zhai, X.; Ren, R.; Hu, J. Total absorption of light in monolayer transition-metal dichalcogenides by critical coupling. *Opt. Express* **2017**, *25*, 31612–31621. [CrossRef] [PubMed]
64. Frindt, R.F. The optical properties of single crystals of WSe_2 and $MoTe_2$. *J. Phys. Chem. Solids* **1963**, *24*, 1107–1108. [CrossRef]

 *nanomaterials*

Article

Two Dimensional β-InSe with Layer-Dependent Properties: Band Alignment, Work Function and Optical Properties

David K. Sang [1,2], Huide Wang [1], Meng Qiu [1], Rui Cao [1], Zhinan Guo [1,*], Jinlai Zhao [1,2], Yu Li [2,*], Quanlan Xiao [1], Dianyuan Fan [1] and Han Zhang [1,*]

1 Shenzhen Key Laboratory of Two Dimensional Materials and Devices, Shenzhen Engineering Laboratory of Phosphorene and Optoelectronics, International Collaborative Laboratory of 2D Materials for Optoelectronics Science and Technology, College of Optoelectronic Engineering, Shenzhen University, Shenzhen 518060, China; dks@szu.edu.cn (D.K.S.); wanghuide@szu.edu.cn (H.W.); qiumeng@szu.edu.cn (M.Q.); caorui@szu.edu.cn (R.C.); zhaojl@szu.edu.cn (J.Z.); xiaoql@szu.edu.cn (Q.X.); fandy@cae.cn (D.F.)

2 College of Materials Science and Engineering, Shenzhen University, Shenzhen Key Laboratory of Special Functional Materials, Shenzhen 518060, China

* Correspondence: guozhinan@szu.edu.cn (Z.G.); liyu@szu.edu.cn (Y.L.); hzhang@szu.edu.cn (H.Z.)

Received: 23 November 2018; Accepted: 27 December 2018; Published: 9 January 2019

Abstract: Density functional theory calculations of the layer (L)-dependent electronic band structure, work function and optical properties of β-InSe have been reported. Owing to the quantum size effects (QSEs) in β-InSe, the band structures exhibit direct-to-indirect transitions from bulk β-InSe to few-layer β-InSe. The work functions decrease monotonically from 5.22 eV (1 L) to 5.0 eV (6 L) and then remain constant at 4.99 eV for 7 L and 8 L and drop down to 4.77 eV (bulk β-InSe). For optical properties, the imaginary part of the dielectric function has a strong dependence on the thickness variation. Layer control in two-dimensional layered materials provides an effective strategy to modulate the layer-dependent properties which have potential applications in the next-generation high performance electronic and optoelectronic devices.

Keywords: Layer-dependent; Indium Selenide; density functional theory; work function; optical properties

---

1. Introduction

Two dimensional layered materials (2DLM) are kind of materials with layered structure which can be obtained by exfoliation of layered bulk materials due to the weak interlayer binding energies [1]. Besides graphene, hexagonal boron nitride (h-BN) [2], transitional metal dichalcogenides (TMDs) [3,4] and black phosphorus (BP) [5,6], are typical 2DLMs. Because of their atomic-level thickness, quantum confinement gives 2DLMs unusual electronic properties, optical properties and thermal conductivity, in contrast to their bulk counterparts. These novel 2DLMs have opened a new twist in understanding the science and technology in nanomaterials.

Although, not each kind of 2DLMs possesses layer dependent physical properties, thickness is one of the most important parameters and should be perfectly controlled when 2DLMs are used in either fabrication of electronic or optical devices. Monolayer and few-layer TMDs such as MoS_2, WS_2, $MoSe_2$ and WSe_2 usually exhibit a thickness-induced indirect-to-direct band gap [7,8]. The excellent optical properties at monolayer level have gained a lot of attraction in terms of applications in lasers and 2D-light-emitting diodes (LEDs) [9,10]. Monolayer WSe_2 exhibits strong optical absorption in the visible range and good light-to-electricity conversion coefficient of approximately 0.5% [11]. The MoS_2-graphene heterojunction with an optimized thickness ratio manifest an enhanced energy conversion coefficient of up to 1% [12], giving rise to potential applications in solar energy conversion devices [13]. Moreover, MoS_2 and WS_2 with different thicknesses have been studied for their potential use in photodetectors [14], valleytronic [15] and spintronics devices [16]. In this case, accurately predicting the physical properties of 2DLMs with exact layer number and obtaining the rule of the layer dependent properties is very important for directing the design and fabrication of the electronic and optoelectronic devices.

Recently, InSe was added into the family of layered transitional metal monochalcogenide. InSe exhibits high photoresponsivity, excellent electrical properties and nonlinear effect and with these extraordinary properties, InSe has captured a lot of attention in the last few years [17–20]. The greatest milestone is the achievement of liquid-phase exfoliated InSe flakes [21–23], which has provided a platform to investigate exciton physics phenomena and engineered practical devices for novel applications in optoelectronic and nanoelectronic field. Recently, Synthesis, electronic properties, ambient stability and applications of InSe were reported [24]. The InSe flakes are not susceptible to oxidation [25–29], even if the Se vacancies induces chemical reactivity towards water [25]. InSe crystals can exist in three polytypes denoted as β, γ and ε phases. γ-InSe is the mostly studied polytype with ABCABC stacking arrangement. Monolayer and few-layer γ-InSe have been proved to possess high electron mobility in the order of 10^3 cm^2 V^{-1} S^{-1} [30,31], excellent metal contact and moderate band gap range [32], which offer the opportunity for presenting tunable nanodevices [33–36]. Sanchez-Royo et al. previously reported a work on γ-InSe, which shares the same composite as β-InSe but stacking in different arrangement [37] and found that there was a huge increase of electronic band gap by more than 1 eV for a single layer and is in agreement with this work, despite the fact that they were investigating the band structures of γ-InSe using DFT approach as implemented in SIESTA code [38]. However, the intrinsic instability of γ-InSe hinders its practical application either in electronics or optoelectronics [39]. The ε–InSe has ACAC layer arrangement and exhibits indirect band structure with a band gap value of 1.4 eV with high photoresponsivity of 34.7 mA/W in few-layer structure [35]. The β-InSe is of great significant since it has been exfoliated into individual layer with hexagonal structure [40], with different electronic and optical properties compared to the bulk β-InSe and it is the most stable phase of InSe due to the ABAB crystal stacking mode [41]. Monolayer and few-layer β-InSe possesses moderate band gap of 2.4 eV and 1.4 eV, respectively [35], which can be an optimal candidate for use in the broadband optoelectronic devices. Moreover, appreciable shift of valence band maximum (VBM) upon thickness variation could be very important for optimizing the band gap to improve electrons and holes mobility. Although, several studies on β-InSe have been carried out to comprehend the tunable performances of the electronic and optoelectronic devices fabricated from it with different layer number and it is important to understand the intrinsic properties which are layer-dependent, that is, band alignment, work function and optical properties.

In this study, the electronic band structures, work function and optical properties of β-InSe monolayer, few-layer and bulk β-InSe have been investigated by performing first principle calculations as implemented in VASP 5.4 [42]. Electronic band structure, work function and optical properties of β-InSe have been demonstrated to be layer thickness-dependent and this provides a wide range of tunability of band gap and corresponding electronic properties, work function and optical properties. The results from this study would provide guideline to experimentalists in obtaining optimal parameters for the design of nanoelectronic and optoelectronic β-InSe-material-based devices.

2. Methods

First-principles calculations based on the density functional theory (DFT) in generalized gradient approximation (GGA) [43], has been performed with the Perdew-Burke-Ernzerhof (PBE) functional [44] for electron exchange-correlation potentials as implemented in the Vienna ab initio Simulation Package (VASP 5.4). The electron-ion interaction was described by employing the projector augmented wave (PAW) method [45] and the cutoff energy for the plane-wave basis was set to 500 eV. To account for the interlayers interaction in few-layer InSe ($L > 1$), we used van der Waals (vdW) correction proposed by

Grimme (DFT-D2) [46]. The Brillouin zone with a Monkhorst-Package scheme $10 \times 10 \times 1$ for *k*-point grid for sampling during structure optimizations and $16 \times 16 \times 1$ in single point calculations was used. The structures were fully optimized via the conjugated gradient algorithm until the equilibrium configuration of atoms was less than 0.01 eVA^{-1}. The convergence criteria energy of electronic in SCF cycles was set to be 10^{-5} eV. In order to mimic the two-dimensional system and to avoid/or make it negligible interaction between repeated unit cells, a vertical separation vacuum space of at least 16 Å was created in the unit cell in the *z*-direction perpendicular to the 2D surface during all calculations. The Phonon property of monolayer β-InSe was calculated using the Full Brillouin zone method implemented in Phonopy [47]. A 4×4 supercell was constructed to calculate the atomic forces employing VASP 5.4, with electronic convergence set to 10^{-8} eV using the normal (block Davidson algorithm).

To confirm the thermal stability of 2D-InSe monolayer, ab initio molecular dynamics (AIMD) simulations within the framework of NVT ensemble (constant number of particles, volume and temperature) [48] was performed. To observe changes in 2D-InSe monolayer structure at the atomic level in the present equilibrium state, a cell with same length, a = b in the x and y direction was considered, respectively. The monolayer was calculated with $4 \times 4 \times 1$ supercell, with 64 atoms in total. Sampling configuration space was carried out at temperature between 300 K and 1000 K. The valence electrons from $4d^{10}5s^25p^1$ for In atom and $3d^{10}4s^24p^4$ for Se atom orbitals are included.

3. Results and Discussion

3.1. Crystal Structure

Bulk InSe can exist in three polytypes denoted as β, γ and ε phases, which show ABAB, ABCABC and ACAC stacking order, respectively. In this work, we focused on the most energetically favorable β-InSe phase, for which the 2DLM crystallizes to form hexagonal structure stacked in AB order, as is shown in Figure 1. The unit cell of β-InSe comprises a base-centered hexagonal lattice classified in the space group $P6_3/mmc$, D^4_{6h}, number 194. From the experimental work [49], lattice constants are a = 4.05 Å, b = 4.05 Å, c = 16.93 Å.

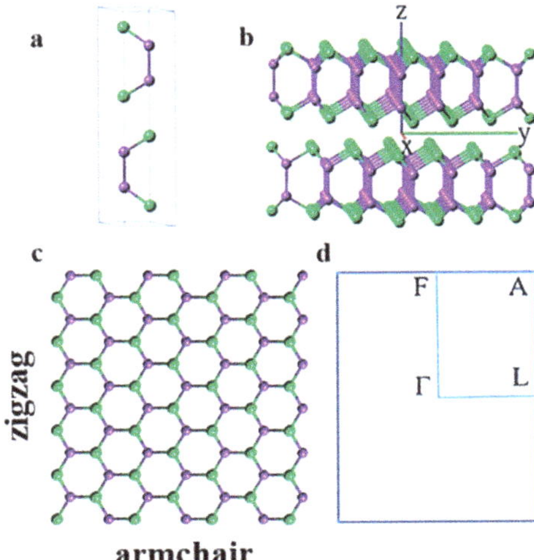

Figure 1. (**a**) Side view of the AB stacking arrangement of atomic structure of a bilayer β-InSe in a unit cell. (**b**) Side view of a bilayer β-InSe. (**c**) Top view of monolayer β-InSe showing armchair and zigzag orientations. (**d**) The 2D-Brillouin zone for monolayer and few-layer β-InSe. The atoms are denoted as In (indigo color) and Se (green color) in the picture.

3.2. Electronic Band Structures of Monolayer and Few-Layer of β-InSe

The study on electronic properties of β-InSe is very significant since it gives insightful description of the system. Owing to the manifestations of the quantum size effects (QSEs) in β-InSe, the nature of the band structure exhibits indirect transitions in monolayer and few-layer. In the scope of this work, $E_F = 0$, is set to be the center of the energy gap. The band gap values of the β-InSe monolayer (1 L), few-layer (2 L to 9 L) and bulk β-InSe within the GGA were tabulated as shown in Table S1 of supplementary information. Calculations in this work showed that, bulk β-InSe, monolayer and few-layer are semiconductors due to the present of band gap in their electronic band structures as shown in Figure S1 of supplementary information. Due to high computation demand from HSE06 functional, only 1 L, 3 L, 5 L and bulk were selected for this study and because GGA-PBE underestimate the band energy value, HSE06 was opted as the best choice because of its accuracy. The calculated band gap values based on HSE06 pseudo-potential for 1 L, 3 L, 5 L and bulk are 2.84 eV, 1.98 eV, 1.84 eV and 1.39 eV, respectively. It is observed that, bulk β-InSe possess direct allowed transition and its band gap energy is comparable with the reported experimental results (1.2~1.30 eV) obtained by use of photoemission electron spectroscopy (ARPES) [50,51] and the β-InSe monolayer (1 L) and few-layer (3 L and 5 L) exhibit robust indirect band gap character as shown in Figure 2. The band gap energy of 1 L (2.84 eV) is comparable with the result from tight-binding model with scissor correction [52]. Magorrian et al. demonstrate that inclusion of spin orbital coupling (SOC) in InSe system has little effect on energy band gap and in this work, it was cross check with monolayer β-InSe and established to be in agreement, as shown in Figure S3b of supplementary information, where the difference is 0.02 eV.

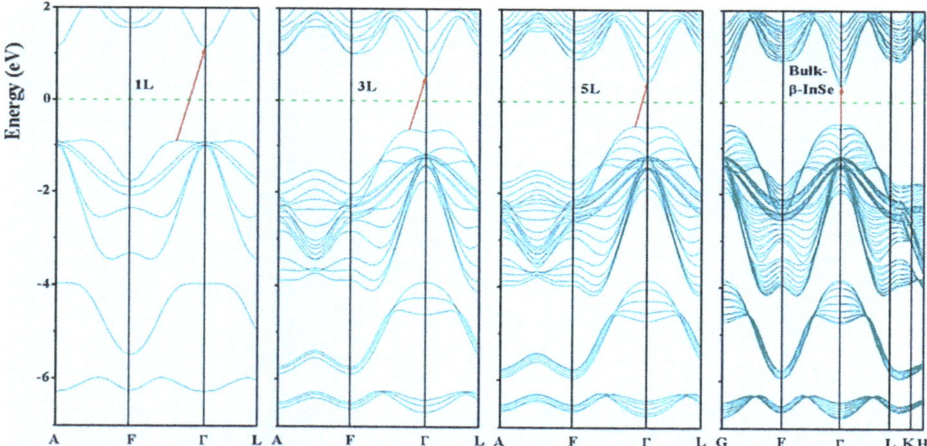

Figure 2. Electronic band structures of β-InSe monolayer (1 L), few-layer (3 L and 5 L) and bulk β-InSe, extracted from HSE06 functional calculations. The green dashed line is Fermi energy level set to 0.0 eV.

The minimum point of conduction band (CBM) appears at the Γ point and is parabolic in nature and the maximum of the first valence band (VBM) appears at a point along the Γ-F direction. Another key character in these band structures is the two peaks at the edge of the valence band maximum (VBM). The two peaks (the so-called sombrero-shape dispersion) at edge of valence band enhances the probability of electron transfer between the valence band and conduction band in monolayer and few-layer, more so for optical conductivity as compare to bulk β-InSe and this fascinating optical response is also highly exhibited by GaS [53].

Moreover, a significant point to note in β-InSe band structures is that, there exist more than one valence band maxima and conduction band minima in almost the same momentum vector, thus electronic allowed transition can take place in these extrema via optical absorption. On the other hand, the band dispersions observed in monolayer, few-layer and bulk are similar and are attributed to uniform crystallinity but the dispersions intensity are quite not the same and this is link to thickness dependency. The symmetric nature of the band dispersion along the high symmetry points of F and Γ in all band structures shows symptomatic isotropic behavior of electronic properties of monolayer and few-layer of β-InSe like graphene, MoS_2 and h-BN layers. The band gap dependence on layer thickness offers tunability of the band gap as well as the corresponding associated electronic properties which is crucial for smart electronic and optoelectronic devices. In order to understand the contribution of different orbitals to the electronic states, calculations of the total DOS and the partial DOS for β-InSe monolayer, few-layer and bulk-β-InSe was performed. The Fermi level in this work is set to $E_F = 0$ eV and is shown by green horizontal dash lines. As shown in Figure 3, the states at the bottom of the VBM have contribution from both p states of In and Se atoms, with little contribution from s states of In and Se atoms. The states forming CBM is mainly derived from the s orbitals of In and p orbitals of Se atoms.

Despite the fact that the valence band have p orbitals from both In and Se atoms, all these orbitals hybridized to form an orbital which is close to the Fermi level energetically. The partial density of states (PDOS) of In and Se atoms shows further evidence in the band hybridization in 1 L, 3 L, 5 L and bulk β-InSe as shown in Figure 3. This hybridization (sp^3) depicts a strong covalent interaction which is a chemical bonding. The trend in the DOS calculation shows that 2D-DOS remained almost constant, giving insights that the electronic states of β-InSe monolayer (1 L), few-layer (2 L to 9 L) have weak dependence on the layer thickness as shown in Figure S2 of supplementary information. The conduction bands of few-layer β-InSe depict finite and almost constant 2D-DOS within a small range of energies due to quantum confinement effects, unlike bulk β-InSe which shows a slight change in DOS within the conduction band region. The valence band in the DOS exhibit a slight shift towards the Fermi level as the layer thickness increase and this shows the effect of layer modulation in β-InSe.

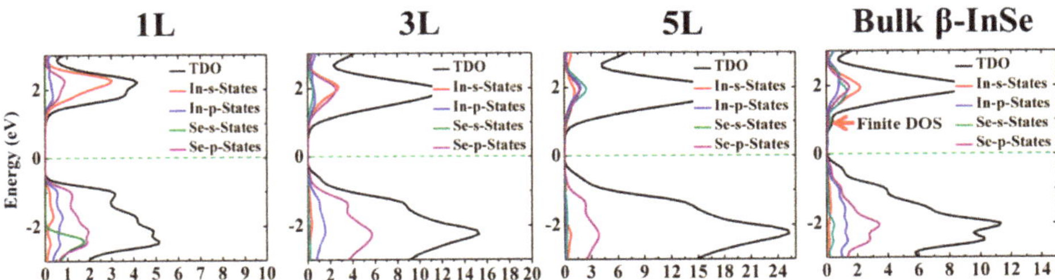

Figure 3. Calculated total density of states (TDOS) and partial density of states (PDOS) of β-InSe monolayer (1 L), few-layer (3 L and 5 L) and bulk β-InSe. The green horizontal dashed line represents Fermi energy level set to be 0 eV.

Since layer control has been used to regulate the electronic properties of semiconductors, also in this study, it gives insights on how the control thickness influences the work function. Work function has been described as energy needed to remove one electron from the system. Therefore, it can be calculated by subtracting the Fermi energy from the electrostatic potential in the middle of the vacuum. From the Figure 4c, it shows that, layer thickness has a strong effect on work functions of β-InSe and hence can be a powerful parameter in controlling the work functions. When the thickness was reduced from bulk to monolayer, the work function of the β-InSe increases monotonically from 4.77 eV to 5.22 eV and this is due to the quantum size effect existing in this atomically thin material (β-InSe monolayer) which make the binding energy of exciton very strong than in bulk β-InSe. The change in work function values as the layer thickness is varied; gives insights on the proper selection of appropriate contacting material with β-InSe and this will offer opportunity for tuning the Schottky barrier to optimal, which in turn improves the carrier mobility.

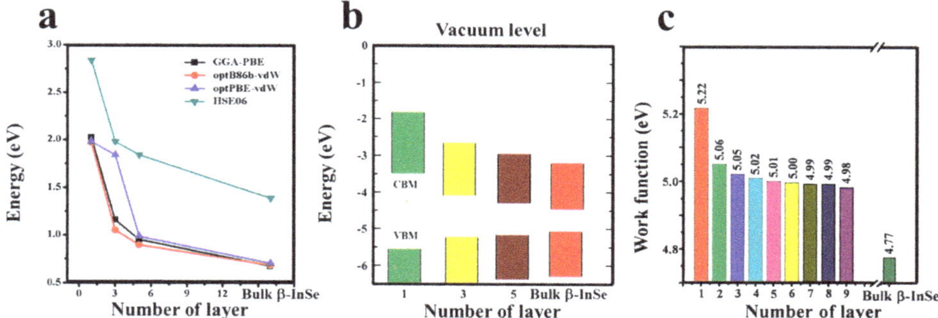

Figure 4. (**a**) Band gap energy of monolayer (1 L), few-layer (3 L and 5 L) and bulk β-InSe as a function of number of layers from various functionals (**b**) Schematic representation of band alignment of monolayer (1 L), few-layer β-InSe (3 L and 5 L) and bulk β-InSe, determined from HSE06 calculations. Vacuum is set as zero for reference. (**c**) Work function of few-layer β-InSe as a function of number of layers.

The β-InSe monolayer and few-layer showed that, the band gap values increase with decrease in the number of layers following a power law as depicted in Figure 4a, showing the variation in the electronic band gap as a function of the layer thickness of β-InSe based on various functionals. The alignments of conduction band minimum (CBM) and valence band maximum (VBM) wavefunctions with respect to the vacuum level are shown in Figure 4b for monolayer (1 L), few-layer β-InSe (3 L and 5 L) and bulk β-InSe, based on hybrid functional calculations. The positions of CBM and VBM for 1 L, 3 L, 5 L and bulk β-InSe are −3.5 eV, −4.0 eV, −4.1 eV, −4.3 eV and −5.5 eV, −3.2 eV, −5.1 eV, −5.0 eV, respectively. The general trend shows the VBM energies shift upwards as the numbers of layers increases and CBM shift downwards. The tuning of the VBM and CBM energies in β-InSe via layer control offers a practical option to optimize the Schottky barrier height, hence improve electron injection efficiency, which leads to more efficient electron mobility across the contact.

To establish the dynamic stability of monolayer InSe, phonon dispersion shown in Figure S4a of supplementary information was calculated within the framework of density functional perturbation theory. The phonon dispersion of monolayer β-InSe showed no imaginary lines in the first Brillouin zone, which is a confirmation that β-InSe is dynamically stable, suggesting the likelihood to be obtained as isolated stable layer.

The performed Molecular dynamics based on direct calculation of the coordinates and velocities of large ensemble of atoms as they evolve over a period of time, gives insightful information on changes in a material's structure at atomic level in the present state of balance. In Figure S4b of supplementary information, confirmed the thermal stability of β-InSe monolayer. Upon the sample was heated at 300 K temperature for 1 ps with a time step of 1 fs, it exhibits structural stability (no structural disorder). Also, it showed that β-InSe can withstand high temperature (1000 K) with no phase change, hence thermodynamically stable. In principal, this is a confirmatory of In-In and In-Se bonds in the β-InSe monolayer that, they are very strong and remained unbroken even when subjected to high temperature and this demonstrate that β-InSe can exist at 300 K and 1000 K. The high thermal stability of β-InSe can be a good material for electronic devices which operates at wide range of temperatures.

3.3. Optical Properties

To advance the materials, in order to gain and retain the recognition in industrial applications, particularly in the optoelectronic field, the optical properties of materials need continuous improvement. The polarizability of the material and polarization directions are the main factors which influence the optical response of the system under study. The polarization of the electric field of incident light (photon) is important and all the optical functions and spectra are calculated along the directions of x and z and are presented relative to hexagonal axis as extraordinary E | | c and ordinary E ⊥ c waves, respectively.

The optical properties of β-InSe were studied via dielectric function. The dielectric function is the summation of real and imaginary parts. The function is as follow:

$$\varepsilon\,(\omega) = \varepsilon_1\,(\omega) + \acute{\iota}\varepsilon_2\,(\omega) \tag{1}$$

This function classifies the material response upon interacting with the electromagnetic spectrum. The real part (ε_1 (ω)) was calculated using the Kramers-Kröningrelation [54]. This relationship is derived from the framework of random phase approximation (RPA) method [55].

$$\varepsilon_1\,(\omega) \;=\; 1 + \frac{2}{\pi}\,p\int_0^{\infty}\frac{\acute{\omega}\varepsilon_2(\acute{\omega})}{\acute{\omega}^2-\omega^2}\,\partial\acute{\omega} \tag{2}$$

where p is the fundamental value (principal value) of this function and other notations are the same as the one in dielectric function.

In Figure 5a, the spectra of real part, ε_1 (ω), in monolayer (1 L) merged at low energy slightly above 0 eV and the spectra in few-layer (3 L and 5 L) set apart at low energy slightly above 0 eV and the separation gradually increase. Compared to bulk β-InSe, the real part is slightly bigger than in few-layer β-InSe. This phenomenon is attributed to the reduced dimensionality and reduced dielectric screening which initiate dramatic change in band gap near the Fermi-level. The energy shift in monolayer and few-layer can provide an opportunity for polarization window be adjusted via layer control, thus β-InSe can be optical linear polarizer. The distinctive optical peaks in spectra remained unchanged as shown in Figure S5 of supplementary information. The real part of the dielectric function is negative at energies between 4–6.5 eV in z-direction (out-of-plane) and 4.5–6.5 eV in x-direction (in-plane), respectively. Only monolayer does not show negative in z direction and this is attributed to the mirror symmetry in the monolayer crystal. The optical function that is, real part (ε_1 (ω)) shows strong dependence on the direction of the electromagnetic wave as shown in Figure 5.

On the other hand, the inter-band momentum allowed transition of electrons is the most important key parameter of the optical spectra characterizer in the semiconductor materials and can be identified in the imaginary part of the dielectric function. To calculate the imaginary part of the dielectric function of β-InSe and bulk β-InSe, need to integrate all possible allowed transitions that is, only from valence to conduction band and to factor in the polarization of the electric field of incident light (photon) which is very significant and therefore, all the optical functions and spectra are calculated along the direction of x and z polarization and are presented relative to hexagonal axis as extraordinary E | | c and ordinary E ⊥ c waves, respectively. Therefore, the imaginary part of the dielectric function is given by the equation:

$$\acute{\iota}\varepsilon_2\,(\omega) \;=\; \frac{4\pi e^2}{m^2\omega^2}\sum_{c,v}\int_B^Z d^3k\,|\langle v_k|p_2|c_k\rangle|^2 \text{ x } \delta\,(E_{ck} - \;E_{vk} - \hbar\omega), \tag{3}$$

where the term $\langle v_k\,|\,p_2\,|\,c_k\rangle$ consist of the occupied and unoccupied states of electrons in the valence and conduction bands. The e and m are the electron charge and mass, respectively. And $\hbar\omega$ is the energy of the incident photon. Since $\varepsilon_{xx} \;=\; \varepsilon_{yy} \neq \; \varepsilon_{zz}$ thus imaginary function $\acute{\iota}\varepsilon_2\,(\omega)$, can be substituted with ε_{xx} and ε_{zz} and the fundamental value p_2, take the form p_{xx} and p_{zz} The imaginary parts of dielectric function of β-InSe monolayer, few-layer and bulk β-InSe against photon energy (eV) are shown in Figure 6.

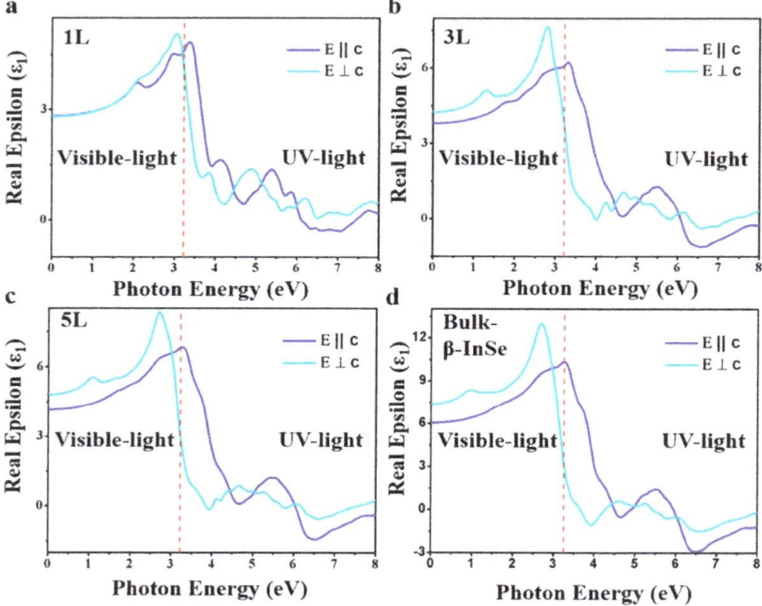

Figure 5. (**a–d**) Calculated real part of the dielectric function a long x and z directions for β-InSe monolayer (1 L), few-layer (3 L and 5 L) and bulk β-InSe.

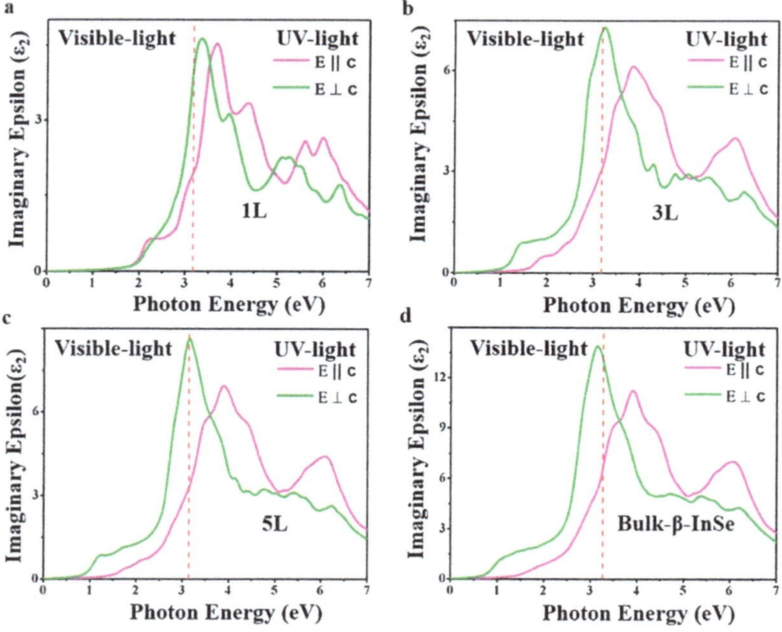

Figure 6. (**a–d**) Calculated imaginary part of the dielectric function a long x and z directions for β-InSe monolayer (1 L), few-layer (3 L and 5 L) and bulk β-InSe.

In Figure 6, the peaks are directly connected to different inter-band transitions in the Brillouin zone and the induced evolution mechanism of the band near the Fermi energy level. In Figure 6, there is a shift in energy from high to low as witness in monolayer (3.3 eV) and bulk β-InSe (3.0 eV), in E $\perp$ c polarization, respectively. This demonstrates that, the control of layer thickness tends to change the frequencies of the absorbed photons and shows that a tunable optical response can be achieved in β-InSe by decreasing layer thickness down to monolayer. The shift into lower energy in β-InSe monolayer and few-layer is due to reduced perpendicular quantum confinement as a result of weak interlayer interaction and this yields weak excitons. Imaginary function ($\acute{\iota}\varepsilon_2$ (ω)) of β-InSe revealed a strong dependence on the direction of the electromagnetic wave as shown in Figure 6.

Other than qualitatively, optical properties can be used quantitatively. The absorptance can increase to a larger value depending on the photon energies range. In this work's calculations, the photon energies range substantially increase in β-InSe monolayer to bulk β-InSe as seen in the peaks intensities and this subsequently increase the absorptance value. Therefore, layer control can be utilized in tuning the absorptance values. In the transparency region of β-InSe, the refractive indices of monolayer (1 L), few-layer (5 L) and bulk β-InSe are found to be 2.35, 3.0 and 3.75 in E $\perp$ c and 2.3, 2.55 and 3.27 in E | | c, respectively, as shown in the Figure S7 and the calculated birefringence (Δn = n_e–n_o) are −0.05 for monolayer (1 L), −0.45 for few-layer (5 L) and −0.48 for bulk β-InSe and all values are negative, indicating that β-InSe is a negative single-axis crystal. The estimated transparency range in the E $\perp$ c polarization direction has been established to be 0.30633 ± 0.04103 and 0.43 μm thickness is required to absorb fully monochromatic light. In the E | | c polarization direction, the transparency has been estimated to be 0.32824 ± 0.01465 and the 0.38 μm thickness is required to absorbed fully monochromatic light.

In this work, the peak intensities of 1 L, 3 L, 5 L and bulk β-InSe that is, heights of fundamental peaks of ($\acute{\iota}\varepsilon_2$ (ω)) in E $\perp$ c polarization have been predicted to be 4.8 (3.3 eV), 7.4 (3.25 eV), 8.5 (3.2 eV) and 14 (3.0 eV) respectively; where the position of the maxima over the photon energy scale is shown in parentheses. As shown in Figure S6, the general trends in peak intensities increase with increase in layer thickness. It reaches maximum in the bulk β-InSe and this is pronounced due to the direct exciton and direct transition. The shape of the peaks in both polarization directions remain constant, demonstrating that, variation in layer thickness do not affect the shape but the photon energies positions and intensities are layer dependent.

Indeed, due to layer control, the band gap increases and decreases depending on the number of layers and this is influenced by the quantum confinement size effects, resulting in the change of orbital overlapping. Thus, the change in the band gap definitely changes the photo-absorption energy that in turn leads to emission of electrons. Therefore, the variations in band gap leads to change in the threshold of imaginary and real part of the dielectric function.

4. Conclusions

We performed a first-principles study on layer dependent electronic band structures, work functions and optical properties of β-InSe, which give advantage in fine tuning the thickness-dependent behavior for possible utilization in the next-generation high-performance electronic and optoelectronic devices. The calculations based on HSE06 functional for 1 L, 3 L, 5 L and bulk has been found to be 2.84 eV, 1.98 eV, 1.84 eV and 1.39 eV, respectively. Bulk β-InSe depicts direct allowed transition and the β-InSe monolayer (1 L) and few-layer (3 L and 5 L) exhibits robust indirect band gaps. Work functions of the β-InSe increases monotonically from 4.77 eV (bulk) to 5.22 eV (monolayer) and understanding the strain effect induced by layer thickness control on work functions of β-InSe is of great important since it offers the achievable route to optimize the Schottky barrier height for future design and developing electronic devices of β-InSe-nanoscale-based devices with superior functionalities. Moreover, optical properties can be efficiently tuned by varying the number of layers as seen in a shift of energy from high (monolayer (3.3 eV) to low (bulk β-InSe (3.0 eV), in E $\perp$ c polarization), which have appeared as exciting technique to tune the optical response in β-InSe. We hope, this study will stimulate further experimental and theoretical studies on this promising β-InSe material and can be extended to other 2D-materials.

Supplementary Materials: The following are available online at http://www.mdpi.com/2079-4991/9/1/82/s1, Figure S1: Electronic band structures of β-InSe monolayer (1 L), few-layer (2 L to 9 L) and bulk β-InSe extracted from GGA-PBE functional calculations, Figure S2a: Calculated total density of states (TDOS) and partial density of states (PDOS) of β-InSe monolayer (1 L) and few layer (2 L to 9 L) and bulk β-InSe based on GGA-PBE functional, Figure S2b: Calculated partial density of states (PDOS) of β-InSe monolayer (1 L) and few layer (2 L and 3 L) and bulk β-InSe, Figure S3: GGA-PBE band gap energies of few-layer of β-InSe as a function of number of layer, Table S3: Tabulation of work function values of β-InSe monolayer (1 L), few-layer (2 L to 9 L) and bulk β-InSe, Figure S4: (a) Calculated phonon band dispersion structure of β-InSe a long high-symmetry direction Γ-Z-M-A-Γ (b) Total potential energy fluctuation of β-InSe monolayer from 500 to 4000 fs during AIMD simulations at the temperature of 300 K, Figure S5: Calculated real part of the dielectric function a long x and z direction for β-InSe monolayer (1 L), few-layer (2 L to 9 L) and bulk β-InSe, Figure S6: Calculated imaginary part of the dielectric function a long x and z directions for β-InSe monolayer (1 L), few-layer (2 L to 9 L) and bulk β-InSe, Figure S7: Calculated refractive index (n) a long x and z directions for β-InSe monolayer (1 L), few-layer (5 L) and bulk β-InSe, Table S1: Tabulation of band gap energy values extracted from different functionals calculations of β-InSe monolayer (1 L), few-layer (3 L and 5 L) and bulk β-InSe, Table S2: Tabulation of band gap energy values extracted from DFT calculations of β-InSe monolayer (1 L), few-layer (2 L to 9 L) and bulk β-InSe based on GGA-PBE.

Author Contributions: D.K.S, Z.G, Y.L and H.Z designed the project; D.K.S performed first principles calculations; D.K.S, Z.G, Y.L, Q.X, D.F, M.Q and H.Z analyzed the data; D.K.S, H.W, R.C and J.Z draft the manuscript. All authors read and approved the final manuscript.

Funding: This research was funded by the National Natural Science Foundation of China (Grant Nos. 61605131, 61435010, 11504239, 61705136).

Acknowledgments: The authors acknowledge financial support from Natural Science Foundation of Guangdong Province for Distinguished Young Scholars 2018B030306038, Natural Science Foundation of Guangdong Province 2016A030310045 and the Science and Technology Innovation Commission of Shenzhen KQTD2015032416270385 and JCYJ20160308092249215.

Conflicts of Interest: The authors declare no conflicts of interest.

References

1. Yoffe, A.D. Layer Compounds. *Annu. Rev. Mater. Sci.* **1973**, *3*, 147–170. [CrossRef]
2. Gorbachev, R.V.; Riaz, I.; Nair, R.R.; Jalil, R.; Britnell, L.; Belle, B.D.; Hill, E.W.; Novoselov, K.S.; Watanabe, K.; Taniguchi, T.; et al. Hunting for Monolayer Boron Nitride: Optical and Raman Signatures. *Small* **2011**, *7*, 465–468. [CrossRef] [PubMed]
3. Mak, K.F.; Lee, C.; Hone, J.; Shan, J.; Heinz, T.F. Atomically thin MoS 2: A new direct-gap semiconductor. *Phys. Rev. Lett.* **2010**, *105*, 136805. [CrossRef] [PubMed]
4. Rai, A.; Movva, H.C.P.; Roy, A.; Taneja, D.; Chowdhury, S.; Banerjee, S.K. Progress in Contact, Doping and Mobility Engineering of MoS2: An Atomically Thin 2D Semiconductor. *Crystals* **2018**, *8*, 316. [CrossRef]
5. Churchill, H.H.; Jarillo-Herrero, P. TWO-DIMENSIONAL CRYSTALS Phosphorus joins the familly. *Nat. Nanotechnol.* **2014**, *9*, 330–331. [CrossRef] [PubMed]
6. Guo, Z.N.; Chen, S.; Wang, Z.Z.; Yang, Z.Y.; Liu, F.; Xu, Y.H.; Wang, J.H.; Yi, Y.; Zhang, H.; Liao, L.; et al. Metal-Ion-Modified Black Phosphorus with Enhanced Stability and Transistor Performance. *Adv. Mater.* **2017**, *29*, 1703811. [CrossRef] [PubMed]
7. Yun, W.S.; Han, S.W.; Hong, S.C.; Kim, I.G.; Lee, J.D. Thickness and strain effects on electronic structures of transition metal dichalcogenides: MX_2 semiconductors (M = Mo, W.; X = S, Se, Te). *Phys. Rev. B* **2012**, *85*, 033305. [CrossRef]
8. Ding, Y.; Wang, Y.; Ni, J.; Shi, L.; Shi, S.; Tang, W. First principles study of structural, vibrational and electronic properties of graphene-like MX_2 (M = Mo, Nb, W, Te; X = S, Se, Te) monolayers. *Phys. B Condens. Matter* **2011**, *406*, 2254–2260. [CrossRef]
9. Baugher, B.W.H.; Churchill, H.O.H.; Yang, Y.; Jarillo-Herrero, P. Optoelectronic devices based on electrically tunable p-n diodes in a monolayer dichalcogenide. *Nat. Nanotechnol.* **2014**, *9*, 262–267. [CrossRef]
10. Ross, J.S.; Klement, P.; Jones, A.M.; Ghimire, N.J.; Yan, J.; Mandrus, D.G.; Taniguchi, T.; Watanabe, K.; Kitamura, K.; Yao, W.; et al. Electrically tunable excitonic light-emitting diodes based on monolayer WSe2 p-n junctions. *Nat. Nanotechnol.* **2014**, *9*, 268–272. [CrossRef]

11. Pospischil, A.; Furchi, M.M.; Mueller, T. Solar-energy conversion and light emission in an atomic monolayer p-n diode. *Nat. Nanotechnol.* **2014**, *9*, 257–261. [CrossRef] [PubMed]
12. Bernardi, M.; Palummo, M.; Grossman, J.C. Extraordinary Sunlight Absorption and One Nanometer Thick Photovoltaics Using Two-Dimensional Monolayer Materials. *Nano Lett.* **2013**, *13*, 3664–3670. [CrossRef] [PubMed]
13. Miao, M.-s.; Botana, J.; Zurek, E.; Hu, T.; Liu, J.; Yang, W. Electron Counting and a Large Family of Two-Dimensional Semiconductors. *Chem. Mater.* **2016**, *28*, 1994–1999. [CrossRef]
14. Gourmelon, E.; Lignier, O.; Hadouda, H.; Couturier, G.; Bernede, J.C.; Tedd, J.; Pouzet, J.; Salardenne, J. MS2 (M = W, Mo) Photosensitive thin films for solar cells. *Solar Energy Mater. Solar Cells* **1997**, *46*, 115–121. [CrossRef]
15. Xiao, D.; Liu, G.-B.; Feng, W.; Xu, X.; Yao, W. Coupled spin and valley physics in monolayers of MoS 2 and other group-VI dichalcogenides. *Phys. Rev. Lett.* **2012**, *108*, 196802. [CrossRef]
16. Zhu, Z.Y.; Cheng, Y.C.; Schwingenschloegl, U. Giant spin-orbit-induced spin splitting in two-dimensional transition-metal dichalcogenide semiconductors. *Phys. Rev. B* **2011**, *84*, 153402. [CrossRef]
17. Lei, S.D.; Wen, F.F.; Ge, L.H.; Najmaei, S.; George, A.; Gong, Y.J.; Gao, W.L.; Jin, Z.H.; Li, B.; Lou, J.; et al. An Atomically Layered InSe Avalanche Photodetector. *Nano Lett.* **2015**, *15*, 3048–3055. [CrossRef] [PubMed]
18. Camara, M.O.D.; Mauger, A.; Devos, I. Electronic structure of the layer compounds GaSe and InSe in a tight-binding approach. *Phys. Rev. B* **2002**, *65*, 125206. [CrossRef]
19. Huang, W.; Gan, L.; Li, H.; Ma, Y.; Zhai, T. 2D layered group IIIA metal chalcogenides: Synthesis, properties and applications in electronics and optoelectronics. *CrystEngComm* **2016**, *18*, 3968–3984. [CrossRef]
20. Tamalampudi, S.R.; Lu, Y.-Y.; Kumar, R.U.; Sankar, R.; Liao, C.-D.; Moorthy, K.B.; Cheng, C.-H.; Chou, F.C.; Chen, Y.-T. High Performance and Bendable Few-Layered InSe Photodetectors with Broad Spectral Response. *Nano Lett.* **2014**, *14*, 2800–2806. [CrossRef] [PubMed]
21. Kang, J.; Wells, S.A.; Sangwan, V.K.; Lam, D.; Liu, X.; Jan, L.; Sofer, Z.; Hersam, M.C. Solution-Based Processing of Optoelectronically Active Indium Selenide. *Adv. Mater.* **2018**, *30*, 1802990. [CrossRef] [PubMed]
22. Petroni, E.; Lago, E.; Bellani, S.; Boukhvalov, D.W.; Politano, A.; Gurbulak, B.; Duman, S.; Prato, M.; Gentiluomo, S.; Oropesa-Nunez, R.; et al. Liquid-Phase Exfoliated Indium-Selenide Flakes and Their Application in Hydrogen Evolution Reaction. *Small* **2018**, *14*, 1800749. [CrossRef]
23. Li, Z.; Qiao, H.; Guo, Z.; Ren, X.; Huang, Z.; Qi, X.; Dhanabalan, S.C.; Ponraj, J.S.; Zhang, D.; Li, J.; et al. High-Performance Photo-Electrochemical Photodetector Based on Liquid-Exfoliated Few-Layered InSe Nanosheets with Enhanced Stability. *Adv. Funct. Mater.* **2018**, *28*, 1705237. [CrossRef]
24. Boukhvalov, D.W.; Gurbulak, B.; Duman, S.; Wang, L.; Politano, A.; Caputi, L.S.; Chiarello, G.; Cupolillo, A. The Advent of Indium Selenide: Synthesis, Electronic Properties, Ambient Stability and Applications. *Nanomaterials* **2017**, *7*, 372. [CrossRef] [PubMed]
25. Politano, A.; Chiarello, G.; Samnakay, R.; Liu, G.; Gurbulak, B.; Duman, S.; Balandin, A.A.; Boukhvalov, D.W. The influence of chemical reactivity of surface defects on ambient-stable InSe-based nanodevices. *Nanoscale* **2016**, *8*, 8474–8479. [CrossRef]
26. Xiao, K.J.; Carvalho, A.; Neto, A.H.C. Defects and oxidation resilience in InSe. *Phys. Rev. B* **2017**, *96*, 054112. [CrossRef]
27. Peng, Q.; Xiong, R.; Sa, B.; Zhou, J.; Wen, C.; Wu, B.; Anpo, M.; Sun, Z. Computational mining of photocatalysts for water splitting hydrogen production: Two-dimensional InSe-family monolayers. *Catal. Sci. Technol.* **2017**, *7*, 2744–2752. [CrossRef]
28. Ho, P.-H.; Chang, Y.-R.; Chu, Y.-C.; Li, M.-K.; Tsai, C.-A.; Wang, W.-H.; Ho, C.-H.; Chen, C.-W.; Chiu, P.-W. High-Mobility InSe Transistors: The Role of Surface Oxides. *Acs Nano* **2017**, *11*, 7362–7370. [CrossRef]
29. Cai, Y.; Zhang, G.; Zhang, Y.-W. Charge Transfer and Functionalization of Monolayer InSe by Physisorption of Small Molecules for Gas Sensing. *J. Phys. Chem. C* **2017**, *121*, 10182–10193. [CrossRef]
30. Sucharitakul, S.; Goble, N.J.; Kumar, U.R.; Sankar, R.; Bogorad, Z.A.; Chou, F.-C.; Chen, Y.-T.; Gao, X.P. Intrinsic electron mobility exceeding 103 cm2/(V s) in multilayer InSe FETs. *Nano Lett.* **2015**, *15*, 3815–3819. [CrossRef]
31. Feng, W.; Zheng, W.; Cao, W.; Hu, P. Back Gated Multilayer InSe Transistors with Enhanced Carrier Mobilities via the Suppression of Carrier Scattering from a Dielectric Interface. *Adv. Mater.* **2014**, *26*, 6587–6593. [CrossRef] [PubMed]

32. Xu, K.; Yin, L.; Huang, Y.; Shifa, T.A.; Chu, J.; Wang, F.; Cheng, R.; Wang, Z.; He, J. Synthesis, properties and applications of 2D layered M III X VI (M = Ga, In; X = S, Se, Te) materials. *Nanoscale* **2016**, *8*, 16802–16818. [CrossRef] [PubMed]
33. Balakrishnan, N.; Kudrynskyi, Z.R.; Fay, M.W.; Mudd, G.W.; Svatek, S.A.; Makarovsky, O.; Kovalyuk, Z.D.; Eaves, L.; Beton, P.H.; Patanè, A. Room temperature electroluminescence from mechanically formed van der Waals III–VI homojunctions and heterojunctions. *Adv. Opt. Mater.* **2014**, *2*, 1064–1069. [CrossRef]
34. -Mudd, G.W.; Svatek, S.A.; Hague, L.; Makarovsky, O.; Kudrynskyi, Z.R.; Mellor, C.J.; Beton, P.H.; Eaves, L.; Novoselov, K.S.; Kovalyuk, Z.D. High Broad-Band Photoresponsivity of Mechanically Formed InSe–Graphene van der Waals Heterostructures. *Adv. Mater.* **2015**, *27*, 3760–3766. [CrossRef] [PubMed]
35. Lei, S.; Ge, L.; Najmaei, S.; George, A.; Kappera, R.; Lou, J.; Chhowalla, M.; Yamaguchi, H.; Gupta, G.; Vajtai, R.; et al. Evolution of the Electronic Band Structure and Efficient Photo-Detection in Atomic Layers of InSe. *Acs Nano* **2014**, *8*, 1263–1272. [CrossRef] [PubMed]
36. Mudd, G.W.; Svatek, S.A.; Ren, T.; Patane, A.; Makarovsky, O.; Eaves, L.; Beton, P.H.; Kovalyuk, Z.D.; Lashkarev, G.V.; Kudrynskyi, Z.R.; et al. Tuning the Bandgap of Exfoliated InSe Nanosheets by Quantum Confinement. *Adv. Mater.* **2013**, *25*, 5714–5718. [CrossRef] [PubMed]
37. Sanchez-Royo, J.F.; Munoz-Matutano, G.; Brotons-Gisbert, M.; Martinez-Pastor, J.P.; Segura, A.; Cantarero, A.; Mata, R.; Canet-Ferrer, J.; Tobias, G.; Canadell, E.; et al. Electronic structure, optical properties, and lattice dynamics in atomically thin indium selenide flakes. *Nano Res.* **2014**, *7*, 1556–1568. [CrossRef]
38. Sanchez-Portal, D.; Ordejon, P.; Canadell, E. Computing the properties of materials from first principles with SIESTA. In *Principles and Applications of Density in Inorganic Chemistry II.*; Kaltsoyannis, N., McGrady, J.E., Eds.; Springer: Berlin, Heidelberg, 2004; Volume 113, pp. 103–170.
39. Bandurin, D.A.; Tyurnina, A.V.; Yu, G.L.; Mishchenko, A.; Zolyomi, V.; Morozov, S.V.; Kumar, R.K.; Gorbachev, R.V.; Kudrynskyi, Z.R.; Pezzini, S.; et al. High electron mobility, quantum Hall effect and anomalous optical response in atomically thin InSe. *Nat. Nanotechnol.* **2017**, *12*, 223. [CrossRef] [PubMed]
40. Debbichi, L.; Eriksson, O.; Lebegue, S. Two-Dimensional Indium Selenides Compounds: An Ab Initio Study. *J. Phys. Chem. Lett.* **2015**, *6*, 3098–3103. [CrossRef]
41. Wang, J.-J.; Wang, Y.-Q.; Cao, F.-F.; Guo, Y.-G.; Wan, L.-J. Synthesis of Monodispersed Wurtzite Structure CuInSe2 Nanocrystals and Their Application in High-Performance Organic-Inorganic Hybrid Photodetectors. *J. Am. Chem. Soc.* **2010**, *132*, 12218–12221. [CrossRef]
42. Kresse, G.; Hafner, J. AB INITIO MOLECULAR-DYNAMICS FOR LIQUID-METALS. *Phys. Rev. B* **1993**, *47*, 558–561. [CrossRef]
43. Zolyomi, V.; Drummond, N.D.; Fal'ko, V.I. Electrons and phonons in single layers of hexagonal indium chalcogenides from ab initio calculations. *Phys. Rev. B* **2014**, *89*, 205416. [CrossRef]
44. Perdew, J.P.; Burke, K.; Ernzerhof, M. Generalized gradient approximation made simple. *Phys. Rev. Lett.* **1996**, *77*, 3865–3868. [CrossRef] [PubMed]
45. Kresse, G.; Joubert, D. From ultrasoft pseudopotentials to the projector augmented-wave method. *Phys. Rev. B* **1999**, *59*, 1758–1775. [CrossRef]
46. Grimme, S. Accurate description of van der Waals complexes by density functional theory including empirical corrections. *J. Comput. Chem.* **2004**, *25*, 1463–1473. [CrossRef] [PubMed]
47. Togo, A.; Tanaka, I. First principles phonon calculations in materials science. *Scr. Mater.* **2015**, *108*, 1–5. [CrossRef]
48. Baroni, S.; de Gironcoli, S.; Dal Corso, A.; Giannozzi, P. Phonons and related crystal properties from density-functional perturbation theory. *Rev. Mod. Phys.* **2001**, *73*, 515–562. [CrossRef]
49. Chen, Z.; Biscaras, J.; Shukla, A. A high performance graphene/few-layer InSe photo-detector. *Nanoscale* **2015**, *7*, 5981–5986. [CrossRef] [PubMed]
50. Politano, A.; Campi, D.; Cattelan, M.; Ben Amara, I.; Jaziri, S.; Mazzotti, A.; Barinov, A.; Gurbulak, B.; Duman, S.; Agnoli, S.; et al. Indium selenide: An insight into electronic band structure and surface excitations. *Sci. Rep.* **2017**, *7*, 3445. [CrossRef]
51. Kibirev, I.A.; Matetskiy, A.V.; Zotov, A.V.; Saranin, A.A. Thickness-dependent transition of the valence band shape from parabolic to Mexican-hat-like in the MBE grown InSe ultrathin films. *Appl. Phys. Lett.* **2018**, *112*, 191602. [CrossRef]
52. Magorrian, S.J.; Zolyomi, V.; Fal'ko, V.I. Electronic and optical properties of two-dimensional InSe from a DFT-parametrized tight-binding model. *Phys. Rev. B* **2016**, *94*, 245431. [CrossRef]

53. Hu, P.; Wang, L.; Yoon, M.; Zhang, J.; Feng, W.; Wang, X.; Wen, Z.; Idrobo, J.C.; Miyamoto, Y.; Geohegan, D.B. Highly responsive ultrathin GaS nanosheet photodetectors on rigid and flexible substrates. *Nano Lett.* **2013**, *13*, 1649–1654. [CrossRef] [PubMed]
54. Jalilian, J.; Safari, M. Electronic and optical properties of alpha-InX (X = S, Se and Te) monolayer: Under strain conditions. *Phys. Lett. A* **2017**, *381*, 1313–1320. [CrossRef]
55. Ehrenreich, H.; Cohen, M.H. SELF-CONSISTENT FIELD APPROACH TO THE MANY-ELECTRON PROBLEM. *Phys. Rev.* **1959**, *115*, 786–790. [CrossRef]

Article

Designing a Novel Monolayer β-CSe for High Performance Photovoltaic Device: An Isoelectronic Counterpart of Blue Phosphorene

Qiang Zhang [1,*], Yajuan Feng [1], Xuanyu Chen [1], Weiwei Zhang [1], Lu Wu [2] and Yuexia Wang [1]

[1] Key Laboratory of Nuclear Physics and Ion-beam Application (MOH), Institute of Modern Physics, Fudan University, Shanghai 200433, China; 14110200002@fudan.edu.cn (Y.F.); 18210200006@fudan.edu.cn (X.C.); 17110200015@fudan.edu.cn (W.Z.); yaowang@fudan.edu.cn (Y.W.)

[2] The First Sub–Institute, Nuclear Power Institute of China, Chengdu 610005, China; lulurichard.student@sina.com

* Correspondence: qiangzhang16@fudan.edu.cn; Tel.: +86-21-5566-4147

Received: 21 February 2019; Accepted: 6 April 2019; Published: 11 April 2019

Abstract: Using the first-principles method, an unmanufactured structure of blue-phosphorus-like monolayer CSe (β-CSe) was predicted to be stable. Slightly anisotropic mechanical characteristics in β-CSe sheet were discovered: it can endure an ultimate stress of 5.6 N/m at 0.1 along an armchair direction, and 5.9 N/m at 0.14 along a zigzag direction. A strain-sensitive transport direction was found in β-CSe, since β-CSe, as an isoelectronic counterpart of blue phosphorene (β-P), also possesses a wide indirect bandgap that is sensitive to the in-plane strain, and its carrier effective mass is strain-dependent. Its indirect bandgap character is robust, except that armchair-dominant strain can drive the indirect-direct transition. We designed a heterojunction by the β-CSe sheet covering α-CSe sheet. The band alignment of the α-CSe/β-CSe interface is a type-II van der Waals *p-n* heterojunction. An appreciable built-in electric field across the interface, which is caused by the charges transfering from β-CSe slab to α-CSe, renders energy bands bending, and it makes photo-generated carriers spatially well-separated. Accordingly, as a metal-free photocatalyst, α-CSe/β-CSe heterojunction was endued an enhanced solar-driven redox ability for photocatalytic water splitting via lessening the electron-hole-pair recombination. This study provides a fundamental insight regarding the designing of the novel structural phase for high-performance light-emitting devices, and it bodes well for application in photocatalysis.

Keywords: density functional theory; mechanical behaviors; electronic properties; type-II heterostructure; photocatalytic properties

1. Introduction

Two-dimensional (2D) families spark tremendous research enthusiasm that is rooted in their exceptional and superior properties [1,2], which is governed by special structure characteristics and quantum size effects, and promising applications in cutting-edge optoelectronic and photonic devices [3,4], supercapacitors [5–7], lithium-ion battery [8–10], efficient ultra-violet photodetector [11], photocatalysis [12–14], field effect transistors [15,16], superior gas sensor [17], and so on. These families, which include semi-metallic graphene (G), semiconducting transition metal dichalcogenides (TMDs), black phosphorene (α-P), and insulating hexagonal boron nitride (*h*-BN), have revealed extensive extraordinary performances. Graphene, which is a zero bandgap semimetal, was reported to possess exceptionally high mobility of carriers and splendid mechanical strength [18–20]. MoS_2, as a typical representative of TMDs, possesses an optimal optical direct band gap [21], high on/off ratios [22], and superior self-healing performances in air [23]. α-P, an atomic ultrathin sheet, which maintains an appropriate direct band gap, was revealed to present extremely high carrier mobility and rather excellent flexibility regarding adjusting electronic and photocatalytic properties by layers stacking and strain engineering [24,25]. Insulating *h*-BN has a broad bandgap of ~6 eV and no dangling bonds, and it can act as an insulating slab and an excellent substrate [26]. Recent progress reported that two vertical heterostructures, including G/α-P and h-BN/α-P, inherit the merits of α-P, i.e., the direct bandgap and linear dichroism, and further improve the properties of α-P concerning easy oxidation when exposed to air [27].

Encouraged by those achievements that were attached to the aforementioned star 2D materials, numerous group IV–VI sheets have been catching people's interest [28–31]. Kamal et al. predicted binary isoelectronic counterparts of α-P and blue phosphorene (β-P), which have almost identical stability in geometric configurations [29]. Among these group IV-VI sheets, several 2D isoelectronic counterparts of α-P, such as α-SnS [32] and α-SnSe monolayers [33], have been manufactured via the exfoliation of their layered bulk whose individual atomic layers are cohered together through weak van der Waals (vdW) force. Indirect-direct transitions that were tuned by stacking coupling for α-SnS [30], or by interface engineering for α-SnSe [34], which can overcome the electron transition obstacle, are now achieved. Very recently, it was reported that monolayer α-CSe, which is an isoelectronic counterpart of α-P with direct bandgap, is highly sensitive to ultraviolet-light [31]. To date, most scientific attentions were thrown at α-phases of group IV-VI. Fortunately, β-P, an allotrope of α-P, has been theoretically predicted [35] and experimentally manufactured [36,37], which aroused the investigations concerning β-P systems, since their α-phases present many novel properties, such as strong in-plane polarization in α-SiS [38] and the high thermoelectric figure of merit in α-SnSe [39]. Currently, only sporadic β-group IV-VI sheets called researches' attention, such as β-SiS [38] and β-SnSe [39]. These made us wonder whether another honeycomb covalent network of C and Se atoms naturally exists, how the mechanical and optoelectronic behaviors perform, and whether it is superior to α-CSe?

Through the calculations using density functional theory (DFT), we identified a novel ultra-thin stable sheet of carbon selenide (β-CSe), which is an isoelectronic counterpart of β-P that also has a wide indirect bandgap. The calculations showed that β-CSe exhibits slightly anisotropic mechanical characteristics. Its bandgap and band-edges curvature are sensitive to in-plane strain, causing the carrier effective mass to be strain-dependent. A vertical heterostructure (α-CSe/β-CSe) based on α-CSe and β-CSe sheets can be constructed at a very small energy penalty, just like the α-P/β-P heterostructure [40]. Its band alignment belongs to that of a type-II van der Waals *p-n* heterojunction. Thus, the electrons and holes are spatially well-separated and distributed in the two layers, respectively. Our results provide a fundamental insight of designing novel structural phase for high-performance photovoltaic devices, and highlight its promising application in photocatalysis.

2. Methods

Calculations were performed in the VASP [41,42] code that was based on density functional theory (DFT). The generalized gradient approximations (GGA) [43] that stemmed from Perdew-Burke-Ernzerhof (PBE) [44] as the exchange-correlation potential within the framework of projector augmented wave (PAW) [45] to model the interactions between electrons and ions were used. A vacuum thickness of 20 Å was adopted to inhibit the mendacious interactions between the periodic nanosheets together with plane-wave basis sets of 500 eV energy cutoff. For Brillouin-zone (BZ) integration, Monkhorst-Pack k-meshes with sizes of $31 \times 31 \times 1$ and $5 \times 33 \times 3$ were applied to the sample in the k-space of monolayer β-CSe and the bilayer α-CSe/β-CSe heterostructure, respectively. All of the structures were fully relaxed with a less-than 0.01 eV/Å residual force on each atom. An energy convergence criterion of 10^{-6} eV was set. As the vdW interactions are crucial in predicting the stable heterostructure, a DFT-D2 semiempirical dispersion-correction approach (considering the vdW interactions) was employed in the calculations of the heterostructure [46]. We adopted the Heyd-Scuseria-Ernzerhof-06 (HSE06) [47] hybrid functional to accurately characterize the electronic performance of monolayer and bilayer. The phonon frequencies were garnered by PHONOPY code [48] based on the density functional perturbation theory (DFPT).

3. Results and Discussions

3.1. Geometric and Electronic Structures

β-P, an allotrope of α-P with intriguing properties, has been theoretically and experimentally discovered [35–37], which is a buckled layer, but it is slightly flat with regard to α-P. We substituted for P-P bonds using C-Se bonds in experimentally well-characterized β-P skeleton. As such, the buckled atomic configuration, named β-CSe, was constructed and is shown in Figure 1a. It is packed into a hexagonal crystal lattice with the *P*3*m*1 space group, one C and one Se atoms are organized in its primitive cell, where each C atom is covalently-linked to its three first-nearest-neighbor Se atoms, and vice versa. These atoms are arranged in the form of a two-atom thick layer with an armchair (zigzag) pattern parallel to x (y) axis, as shown in Figure 1a. Table 1 lists the well-optimized structural parameters for the β-CSe and β-P monolayers, which cater to previous results [29,35]. When compared to β-P, the different local bonding preference between C and Se atoms alters the buckled height of monolayer β-CSe. The buckled angle of β-CSe (96.45°) exceeds that of β-P (92.907°); therefore, its buckled height is reduced to 1.044 Å with regard to β-P (1.238 Å) [35].

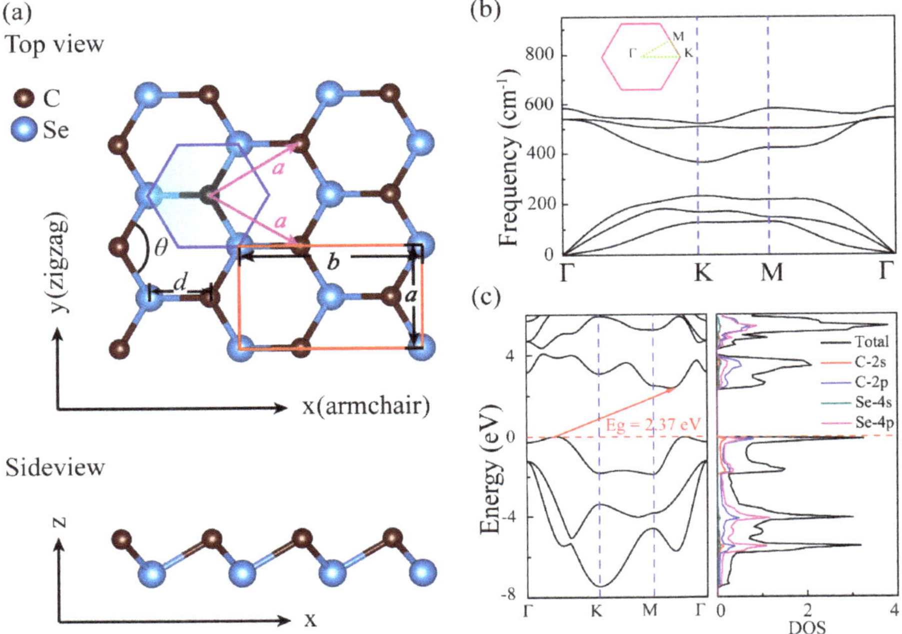

Figure 1. (**a**) The top and side views of monolayer carbon selenide (β-CSe). The shade region represents a primitive cell. A rectangular cell in top view also is marked, and used to calculate stress-strain relationships. (**b**) Phonon band diagram, where the panel represents the high-symmetry k-points in the first BZ of the hexagonal reciprocal unit cell. (**c**) Electronic band diagram, and the total and orbital projected partial density of states for the monolayer β-CSe, at the HSE06 level. The indirect bandgap of the monolayer is guided by the red arrow and the bandgap value is also provided. The Fermi level is set at zero.

Table 1. The cohesive energy and optimized structural parameters of the monolayer β-CSe and blue phosphorene (β-P).

Structure	Space Group	Cohesive Energy (eV/atom)	Lattice Constants (Å)		Bond Length (Å)	Bond Angle (deg)
			a	*b*		
β-CSe	$P3m1$	−3.79	3.065	5.22	2.055	$\theta = 96.45$
β-P	$P3m1$	−5.23	3.28/		2.261	$\theta = 92.907$

The stability of β-CSe is first evaluated by the cohesive energy and the phonon calculations. Like the phosphorene allotropes (α-P and β-P), a cohesive energy of β-CSe (−3.79 eV/atom) slightly deviates from α-CSe (−3.86 eV/atom). This small difference is comparable to the thermal energy at 300 K [49]. Hence, the monolayer β-CSe is predicted to be stable in energy, just as the α-CSe sheet. As shown in Figure 1b, no imaginary phonon modes appear in the first BZ, revealing that its structure has outstanding dynamical stability.

Table 2 lists the gaps of β-CSe and β-P from PBE and HSE06 calculations, which match well with other works [29,35]. Figure 1c shows the electronic structure and total and orbital projected partial densities of states (TDOS and PDOS) of β-CSe from HSE06 calculations. PDOS reveals that those states in the valence band maximum (VBM) and the conduction band minimum (CBM), are principally contributed by p electrons with less weight of s electrons. Specifically, these electronic states approaching the Fermi level have larger contributions from $2p$ electrons of C atoms instead of $4p$ electrons of Se atoms. This is mainly rooted in the fact that C has electronegativity greater than Se [29].

Table 2. The bandgaps and carrier effective mass of monolayer β-CSe and β-P. Indirect bandgap is marked as In in parenthesis.

Material	PBE Gap (type) eV	HSE Gap (type) eV	m^*_h/m_e Zigzag Direction	m^*_h/m_e Armchair Direction	m^*_e/m_e Zigzag Direction	m^*_e/m_e Armchair Direction
β-CSe	1.54 (In)	2.37 (In)	0.718	0.795	0.23	1.027
β-P	1.94 (In)	2.7 (In)	0.588	0.486	0.353	0.794

3.2. Mechanical Properties

Strain engineering provides an efficient method for investigating the stress-strain relation of 2D sheets [50–52]. Specifically, the strain engineering, which is implemented by deliberately imposing mechanical deformation onto an ultrathin sheet, shows that the strain itself potentially serves as a practical tool for investigating the nanostructure response to it. The mechanical behaviors of the monolayer β-CSe under uniaxial strain along the armchair (ε_{xx}) and the zigzag directions (ε_{yy}), and the uniform biaxial strain (equiaxial strain ε_{xy} along ab double axes) are calculated. The engineering strain is denoted as $\varepsilon = (n - n_0)/n_0$, where n and n_0 represent the lattice constants of the strained and equilibrium structures, respectively. The positive ε refers to the extension, while negative ε, contraction. When imposing the uniaxial strain (the lattice constant in the strain axis is fixed), the in-plane unstrained lattice vector is fully relaxed. For the biaxial strain, only atoms positions in the unit cell are fully relaxed.

The defined rectangular unit cell (the red frame) in Figure 1a is adopted to conveniently implement the stress-strain simulation. The evolution of its total energy under axial strains is examined first, as shown in Figure 2a. A potential well can be attained, in which the minimum corresponds to the relaxed structure. The curvature of energy under the armchair strain is smaller than that under the zigzag strain, indicating that it is easily deformed along the armchair direction with respect to the zigzag direction. Figure 2b indicates that the monolayer β-CSe will contract (expand) in the other perpendicular direction, when it is stretched (compressed) in the armchair- or zigzag-directions. The calculated Poisson's ratios are 0.14 under the armchair strain and 0.16 under the zigzag strain. A tiny deviation of Poisson's ratios between the two perpendicular directions indicate the slightly anisotropic nature. The Poisson's ratios are small, being comparable to those of other 2D materials [49], for example, 0.12 for a monolayer C_3N, 0.21 for a monolayer BN, and 0.29 for a monolayer SiC, which are all knitted by sp^3 bonds. In order to depict the corrugation of monolayer β-CSe, Figure 2c shows the dependences of its buckled height on the uniaxial and biaxial tensions, which is distinct from that of α-P [53]. Clearly, Figure 2c shows that the monolayer β-CSe expands (compresses) in the z-direction when it is contracted (stretched) under all of the strain cases, and thus, negative linear Poisson's ratio is not appeared like the monolayer α-P. The evolutions of the buckled height is different in the three strain cases: the maximum of 1.05 Å appears at the armchair strain of $\varepsilon_{xx} = -0.07$; whereas, the maximum appears at the zigzag strain of $\varepsilon_{yy} = -0.20$; the equibiaxial strain remarkably causes the largest variation of the buckled height in the range of 1.4 Å~0 Å. When the equibiaxial strain exceeds $\varepsilon_{xy} = 0.14$, the corrugation rapidly diminishes and vanishes at $\varepsilon_{xy} = 0.18$. It implies a structure phase transition from initially the low-buckled configuration to plane structure. Accordingly, orbital hybridization transfers from the weak sp^3 to sp^2 hybridization.

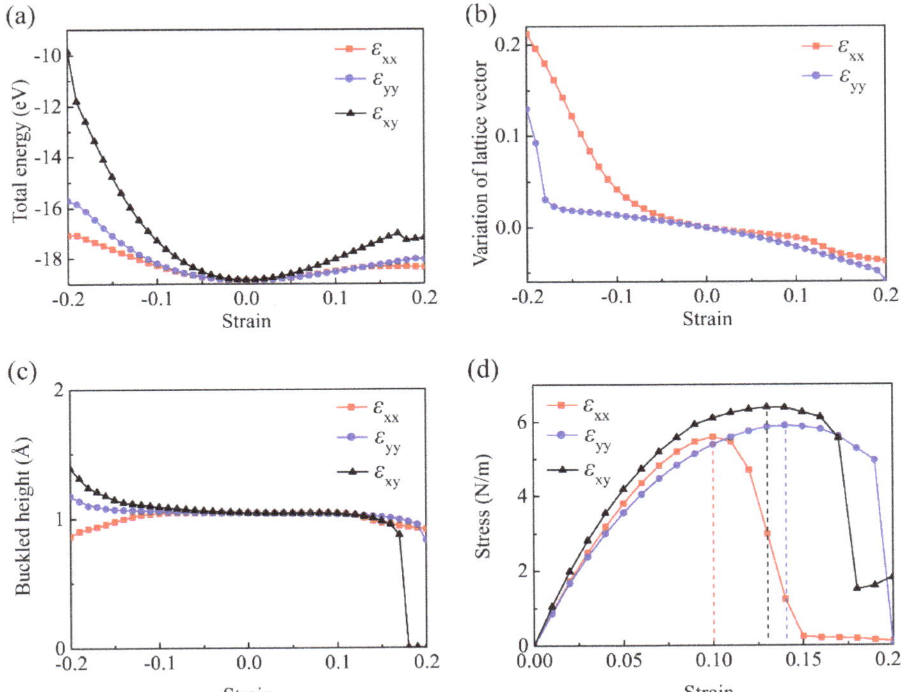

Figure 2. (**a**) Calculated total energy of the monolayer β-CSe under strains. (**b**) Variation of the lattice vector perpendicular to the strain direction. (**c**) Buckled height of the monolayer β-CSe under the uniaxial and biaxial strains. (**d**) Stress-strain relationships of themonolayer β-CSe under three types of strains.

Currently, the interlayer distance of the monolayer β-CSe cannot be experimentally determined. Therefore, the in-plane stress (2D stress per unit length) can be used to characterize the strength of the monolayer β-CSe [50]. To explore its ideal tensile strength, the uniaxial and biaxial tensile strains are exerted on the relaxed monolayer. As depicted in Figure 2d, the stress-strain curve under the armchair tension almost overlaps that of zigzag tension below a loading of 0.03, i.e., β-CSe possesses isotropic in-plane elastic response in this strain interval. As the tension further increases, the stress-strain behavior become nonlinear, and the disparity of elastic response in both orthonormal directions becomes conspicuous. Table 3 summarizes the ideal strengths and critical strains, which reveal the monolayer β-CSe can withstand an ultimate stress of 5.6 N/m along the armchair direction, and 5.9 N/m along the zigzag direction. We found that the ideal tensile strengths are nearly identical, just like graphene [54] and silicene [55], which is attributed to the similar rhombic hexagonal structure. Clearly, it can withstand a tensile strain ultimate of 0.1 in the armchair direction and 0.14 along the zigzag direction. These critical strains are found to be small, especially, a critical strain of 0.1 along the armchair direction (as we know, it is the smallest in extensively atomically studied ultra-thin sheets, excluding borophene [50]). Accordingly, β-CSe becomes an outstanding candidate for brittle materials. In the case of the biaxial tension, this monolayer can withstand a stress up to 6.4 N/m at $\varepsilon_{xy} = 0.13$. Intriguingly, the curve presents an inflexion point with a minimum value under a biaxial strain level of $\varepsilon_{xy} = 0.18$, which is responsible for structure phase transition. The Young's modulus can also be garnered via fitting the initially linear segment of stress-strain curves up to 0.02 for the uniaxial or biaxial strains and Table 3 lists the corresponding values, which reveals that the strength of C-Se bonds may be slightly different between the both orthogonal directions, and it further confirms the anisotropic nature of the monolayer β-CSe.

Table 3. The ideal strengths (f), critical strains (ε_c), Young's modulus and Poisson's ratios of monolayer β-CSe under the three strain styles.

Direction	f (N/m)	ε_c	Young's Modulus (N/m)	Poisson's Ratio
Armchair	5.6	0.10	86.14	0.14
Zigzag	5.90	0.14	83.47	0.16
Biaxial	6.4	0.13	99.15	0.09

3.3. Strain Dependence of the Electronic Structures

Strain engineering has been proved as a commendable avenue for tuning the optoelectronic properties of 2D semiconductors [28,49,56,57]. Consequently, we have carried out calculations regarding the electronic structures of the strained monolayer β-CSe. The electronic structures under ε_{xx} are firstly explored, as shown in Figure 3a. It was found that indirect-to-direct band transition appears under the tension ($\varepsilon_{xx} = 0.16$) or compression ($\varepsilon_{xx} = -0.15$) along the armchair direction. The two direct band gaps reside at non-identical high-symmetry points (Γ for tension and the point between Γ and Y for compression). For the case of the band structures under ε_{yy} or ε_{xy}, as presented in Figure 3b,c, no indirect-to-direct band transition takes place.

Figure 3 illustrates that the positions of VBM and CBM frequently change. The monolayer β-CSe at the equilibrium configuration presented indirect behavior with the CBM positioned in Γ-X and VBM lying in Γ-Y, which correspond to the electronic structure of zero strain in Figure 3. Clearly, all of the strained systems experience a semiconductor-metal transition under larger strains, except for large armchair tensile strain. In what follows, specific discussions are focused on semiconductors. For the case of ε_{xx}, with increasing tension, the CBM changes to Γ-S firstly and then shifts to Γ-Y (including Γ), whereas the VBM always lies in Γ-Y. Similarly, with increasing compression, the CBM always lies in Γ-X (including Γ), while the VBM position mainly depends on the competition between the band edges states (Γ-X and X-S). For ε_{yy}, with an increase of tension, the CBM shifts from Γ-X to Γ, while the VBM position is decided by fierce competition between Γ-X and Γ-Y. On the side of compressive strain, the CBM lies in Γ-S and VBM shifts to Γ from Γ-Y with increasing the compression. In the case of ε_{xy}, the CBM position mainly depends on the competition between the band edges states (Γ-Y and Γ-S), while VBM always appears between Γ-Y.

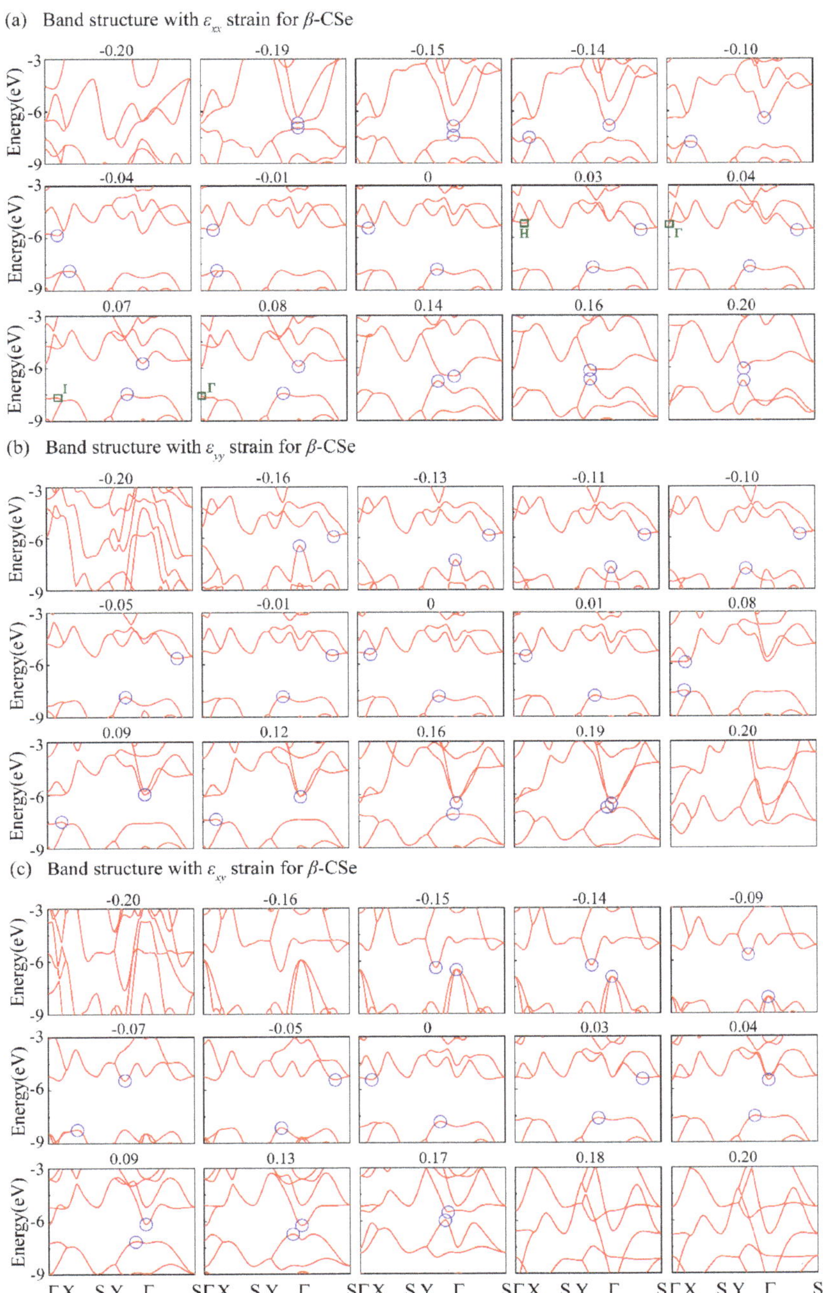

Figure 3. Evaluation of band structures versus armchair strain (**a**), zigzag strain (**b**) and biaxial strain (**c**). The blue circles represent the conduction band minimum (CBM) and valence band maximum (VBM). The energy values are relative to the vacuum level.

3.4. Strain Dependence of the Bandgap

The strain-induced electronic structures evolutions issue in significant changes in its band gap. We plot the strain-induced bandgap as functions of ε_{xx}, ε_{yy}, and ε_{xy}, as captured in Figure 4 to clearly elaborate it. One can see that the bandgap is first diminished versus the increasing of the armchair tension until the tensile strain reaches $\varepsilon_{xx} = 0.13$, after which the bandgap increases. On the side of contraction, it declines nearly monotonously and even vanishes at $\varepsilon_{xx} = -0.20$. Under ε_{yy}, the band gap narrows to be zero with increasing either the tension or compression. When it was subjected to the equiaxial in-plane tensile strain, the band gap non-monotonously descends and even closes at $\varepsilon_{xy} = 0.18$. Upon imposing the compressive strain, the transition location of the band gap from ascension to descension appears at $\varepsilon_{xy} = -0.07$ with increasing the compression, and eventually the bandgap closes. Overall, the armchair-dominant strain issues in an indirect-direct transition despite the indirect characterize being robust. By contrast, the cases are exceedingly distinct for the zigzag and the biaxial strains, where only a transition from semiconductor to metal is observed and the indirect semiconductor nature is well preserved in sizable strain intervals.

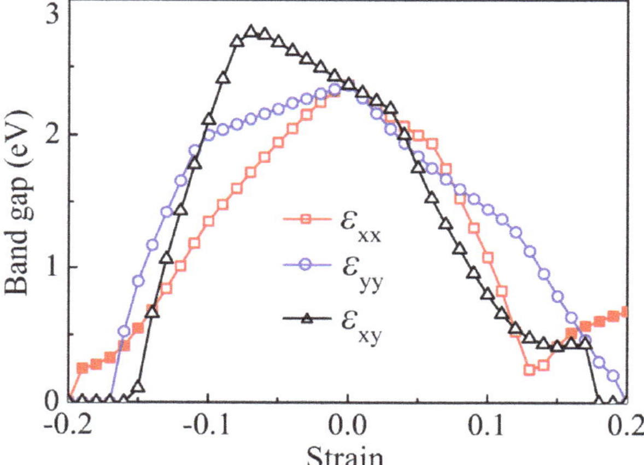

Figure 4. Bandgap of the monolayer β-CSe varies with the uniaxial and biaxial strains. Red line, blue line and black line represent the ε_{xx}-, ε_{yy} -, and ε_{xy} -induced band gap evolutions, respectively. The hollow and solid symbols indicate the indirect and direct electron transitions, respectively.

3.5. The Dependence of Carrier Effective Masse on Strain

The strain in semiconductors can modulate not only the characteristic of electron transitions, band-edges positions, and the band gap, but also the curvature of the electronic band edges. The curvature determines the carrier effective mass m^* : $m^* = \hbar^2(\partial^2 E/\partial K^2)^{-1}$, where $\hbar$, E, and K refers to the reduced Planck constant, energy, and momentum, respectively. Table 2 summarizes the effective masses of the relaxed β-CSe and β-P sheets, among which the results of monolayer β-CSe have not been reported before and the results of monolayer β-P reasonably agree with previous works [58]. Figure 5a,b show the variation of carrier effective mass versus strains in this work. When compared to the zigzag direction, strain exerts a stronger effect on the effective carrier mass of the armchair direction. As for ε_{xx}, the effective electron mass of the armchair direction ($m_{exx}^{armchair}$) presents a sudden drop when the strain exceeds $\varepsilon_{xx} = 0.03$, and the situation reappears for the effective hole mass of the armchair direction ($m_{hxx}^{armchair}$) at $\varepsilon_{xx} = 0.08$. Specifically, the effective electron mass of the armchair direction far outweighs that of the zigzag direction under $\varepsilon_{xx} < 0.04$, which indicates that the preferred transport is along the zigzag direction for electrons in the corresponding strain interval. However, under $\varepsilon_{xx} > 0.04$, the armchair direction becomes the dominant direction for electron transport. For the effective hole mass, there is a transition strain ($\varepsilon_{xx} = 0.07$), causing a sudden transition from $m_{hxx}^{armchair} > m_{hxx}^{zigzag}$ to $m_{hxx}^{armchair} < m_{hxx}^{zigzag}$ when the ε_{xx} exceeds the transition strain. It indicates that the prior direction for hole transport has a sudden change under the tensile strain.

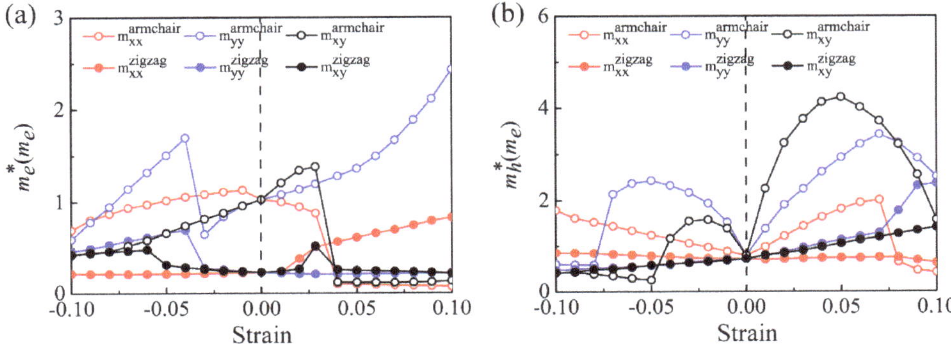

Figure 5. The effective electron (**a**) and hole (**b**) masses as functions of strains (ε_{xx}(red), ε_{yy} (blue), and ε_{xy} (black)). Hollow circles and solid circles represent the armchair and zigzag effective masses, respectively. The carrier effective mass is in unit of the static electron mass (m_e).

ε_{yy} can also remarkably modulate the effective carrier masses, as shown in the blue lines of Figure 5a,b. For the electrons and holes, a non-monotonous dependence of the effective masses on the zigzag strain was conspicuous. $m_{yy}^{armchair} < m_{yy}^{zigzag}$ remains over the whole strain ranges, where $m_{yy}^{armchair}$ is the carrier effective mass of the armchair direction under the zigzag strain and m_{yy}^{zigzag} is the carrier effective mass of the zigzag direction under the zigzag strain. Consequently, the armchair direction undertakes the dominant transport direction for the electrons and holes, which cannot be changed by the strain. This implies that the transport performances of carriers in the monolayer β-CSe preserve mechanical stability.

For the case of ε_{xy}, the biaxial strain provided distinct modulations for the effective carrier mass with regard to the aforementioned uniaxial strain. For electrons, the large anisotropy is decayed and even gradually disappears with increasing compressive biaxial strain, i.e., the favored transport direction has a transition from the single armchair direction to the double directions under $\varepsilon_{xy} = -0.08$. Around $\varepsilon_{xy} = 0.04$, a sudden drop of the effective mass along the armchair direction takes place, which causes a preferential transport direction transition from the zigzag direction to the armchair direction. For the holes, at $\varepsilon_{xy} < 0$, the anisotropy feature of the effective masses is weakened with an increase of the biaxial compression: at $\varepsilon_{xy} = -0.05$, the effective mass has an apparent transition from $m_{hxy}^{armchair} > m_{hxy}^{zigzag}$ to ($m_{hxy}^{armchair}$ and m_{hxy}^{zigzag} represent the hole effective masses of the armchair and zigzag directions under the biaxial strain, respectively); at $\varepsilon_{xy} = -0.09$, the anisotropy of the effective mass disappears ($m_{hxy}^{armchair} = m_{hxy}^{zigzag}$); at $\varepsilon_{xy} > 0$, the holes are always heavier than the electrons. Clearly, the larger anisotropy in the carrier effective mass during load will result in anisotropic carrier mobility, and will give further rise to direction-dependent conductivity. Overall, the sudden and frequent shift about the effective masses of the electrons and the holes leads to competition between two orthonormal directions concerning the preferred transport.

The effective mass shown in Figure 5 is directly related to the electronic structures shown in Figure 3. Particularly, the dramatic shift regarding the effective mass is due to the fierce competition of band-edge extremes. In order to have an in-depth understanding about the dramatic shift, only the armchair strain is taken as a typical reference since the sharp change of effective carrier mass in other strain cases is similar in physical mechanism. Specifically, the dramatically downward change of the effective electron mass in the armchair direction around $\varepsilon_{xx} = 0.04$ in Figure 5a, consults the band structure of Figure 3a along Γ-X (armchair). When compared to the state H in Figure 3a at $\varepsilon_{xx} = 0.03$, the downward shift of state Γ under $\varepsilon_{xx} = 0.04$ obviously strengthens the band dispersion along the armchair direction, and it thus dramatically decreases the effective electron mass at $\varepsilon_{xx} = 0.03$. The armchair CBM is departed from H to Γ when the strain exceeds $\varepsilon_{xx} = 0.03$. Thereby, the calculated effective electron mass from state Γ is smaller, because of strengthened dispersive at the new armchair CBM. Another dramatic downshift in the effective hole mass appears at $\varepsilon_{xx} = 0.08$ in Figure 5b, which is closely related to the band structure of Figure 3a at $\varepsilon_{xx} = 0.08$. Here, the energy of the valence-band state Γ surpasses the state I and then becomes the new armchair VBM. The effective hole mass of armchair direction is now calculated according to this emerging armchair VBM (the state Γ) rather than the state I.

3.6. The Type-II vdW p-n α-CSe/β-CSe Hetrostructure as a Metal-Free Photocatalyst

As a new degree of freedom, which introduces interfacial coupling, is also expected to tailor optoelectronic performance. This stimulates us to propose the α-CSe/β-CSe vdW heterostructure. The unit cell of the proposed vdW nanocomposite was constructed by placing the super cell, which includes 4 × 1 rectangle unit cells of β-CSe on the top of the super cell includes 5 × 1 unit cells of α-CSe. The fully-optimized atomic motif that was obtained from the calculations of DFT+D2 function is shown in the upper (topview) and lower (sideview) panels of Figure 6a. The well-optimized lattice parameters of 5 × 1 super cell of α-CSe and 4 × 1 rectangle super cell of β-CSe are a = 3.048 Å and b = 21.47 Å, and a = 3.057 Å and b = 20.89 Å, respectively. Such a complex has the well-optimized lattice constants of a = 3.051 Å and b = 21.08 Å. Consequently, the overall induced largest mismatch is 1.82% in α-CSe along the b-direction, allowing for one to engineer its optoelectronic properties at a low energy penalty.

The effect of vdW interaction can be authenticated via analyzing the electronic structure. As shown in Figure 6b, the hybrid α-CSe/β-CSe vdW heterojunction has an indirect band gap of 1.4 eV at the HSE06 level, and the VBM is positioned at a non-high-symmetry k-point along Γ-X, whereas the CBM appears at Γ. Obviously, the vdW interaction does reduce the band gap and it reshapes energy band extremum (EBE). In order to uncover the origin of EBE, DOS is plotted, as presented in Figure 6c. The sulfur yellow and magenta (blue areas) in Figure 6c represent the contributions from the relaxed heterostructure and the pure α-CSe (β-CSe) layer, respectively. The results indicate that the CBM are dominated by α-CSe, whereas the VBM is mainly rooted in β-CSe, i.e., the vdW interaction in this hybrid heterojunction changes the positions of EBE in comparison with the free-standing monolayer β-CSe, and forms an atomically sharp type-II vdW heterostructure. In this type-II heterostructure, the photon-generated electron-holes pairs should be separated in space, in the form of electrons and holes that were allocated in different layers. In-depth analysis concerning the partial charge densities of CBM and VBM further supports that this heterostructure belongs to Type-II vdW heterostructure, because the CBM (Figure 6d) and VBM (Figure 6e) are mainly originated from the states of the α-CSe layer and the β-CSe layer, respectively. These separated optically active states in space are equivalent to spontaneously separated carriers that were generated from photon, which helps to improve the solar energy conversion efficiency. The band alignment of α-CSe/β-CSe also supports this conclusion, as follows. To achieve the band alignment between α-CSe and β-CSe, we firstly ascertain the positions of band edges with respect to the vacuum level. The positions of band edges (referred to the vacuum level) in the α-CSe and β-CSe sheets before and after contacting can be obtained by solving the Kohn–Sham equation. More specifically, after α-CSe contacts β-CSe, the bandgap (1.52 eV) of the α-CSe layer is approximately unchanged when compared with that of the isolated one [31], spanning an energy range from −5.86 to −4.34 eV, while the band gap of the β-CSe monolayer apparently reduces to 2.17 eV with regard to that of the isolated monolayer β-CSe, spanning from −5.74 eV to −3.57 eV. Therefore, the CBM and VBM come from different layers, resulting in a typical type-II vdW heterostructure.

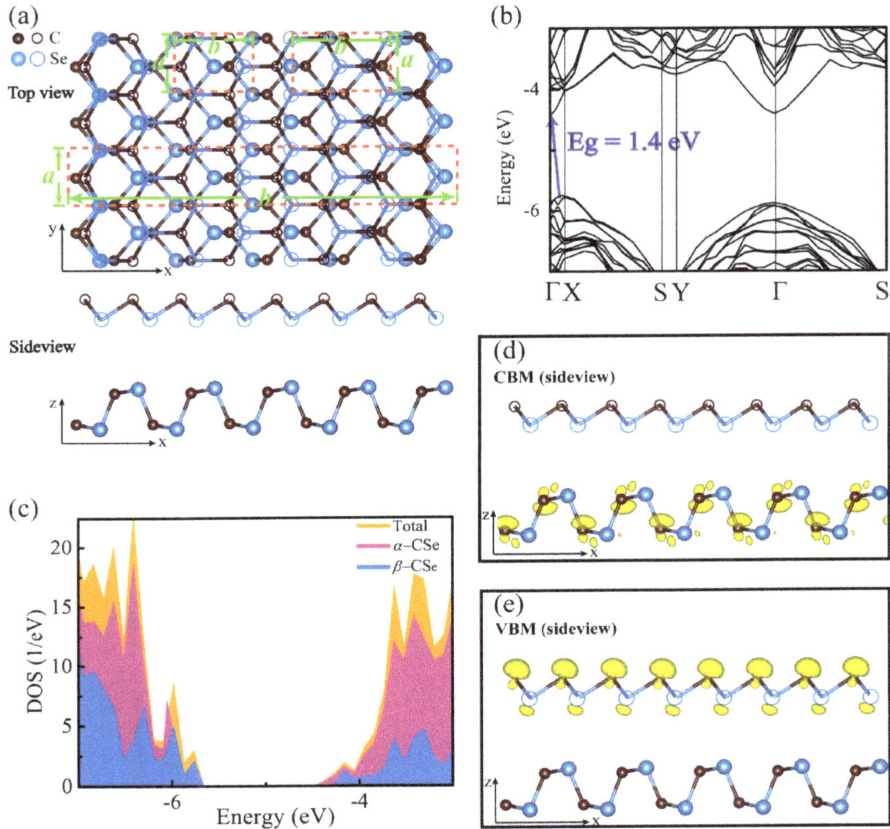

Figure 6. Top- and side-views (**a**) for the α-CSe/β-CSe vdW hetrostructure. The three red dashed frames are the rectangle unit cell of α-CSe, β-CSe and the heterostructure, respectively. The band structure (**b**) and density of states (**c**) refer to vacuum level. The band decomposed charge density of the CBM (**d**) and VBM (**e**) in the α-CSe/β-CSe vdW heterostructure. The value of isosurfaces is 0.03 e/Å^3. The hollow and solid spheres represent the top and bottom layers, respectively.

In order to precisely characterize such vdW coupling between the α-CSe layer and the β-CSe slab, we calculated the planar-averaged charge density difference (CDD), which is defined as $\Delta\rho(z) = \rho_{\alpha-CSe/\beta-CSe} - \rho_{\alpha-CSe} - \rho_{\beta-CSe}$, here $\rho_{\alpha-CSe/\beta-CSe}$,$\rho_{\alpha-CSe}$, and $\rho_{\beta-CSe}$ represent the charge densities of the hybrid heterojunction architecture, the pure α-CSe slab, and the pure β-CSe sheet, respectively. Figure 7a shows the planar-averaged CDD (black line) of the α-CSe/β-CSe heterostructure as a function of the z-axial position. The change of position-dependent $\Delta\rho(z)$ at interface evidences that the β-CSe layer contributes electrons to the α-CSe layer, which induces a slight n-type doping in the α-CSe layer and a p-type doping in the β-CSe layer. To quantify the transfer of charge, the electron-transfer quantity (red line) up to z point can be acquired by $\Delta Q(z) = \int_{-\infty}^{z} \Delta\rho(z')dz'$. The electron gain in the α-CSe layer is 0.028 e, as calculated from the value of $\Delta Q(z)$ at the charge-transfer complex interface. Such small charge transfer reveals a weak interlayer coupling between the α-CSe layer and the β-CSe layer. To further unveil the charge-transfer mechanism, Figure 7c provides the three-dimensional isosurface of the CDD, where the charge accumulation (yellow) and the depletion (cyan) of electrons across the interface are intuitively illustrated. Apparently, the charge density redistributes in the interface region of the heterostructure. The holes accumulate near the β-CSe region, whereas the electrons accumulate in the region near the α-CSe layer. In the formation of a *p*-*n* α-CSe/β-CSe vdW heterostructure, charge density redistributes and the Fermi level is driven to the CBM of α-CSe and VBM of β-CSe after they contact. Meanwhile, the α-CSe layer acts as a role of the electron acceptor, while the β-CSe layer served as an electron donor.

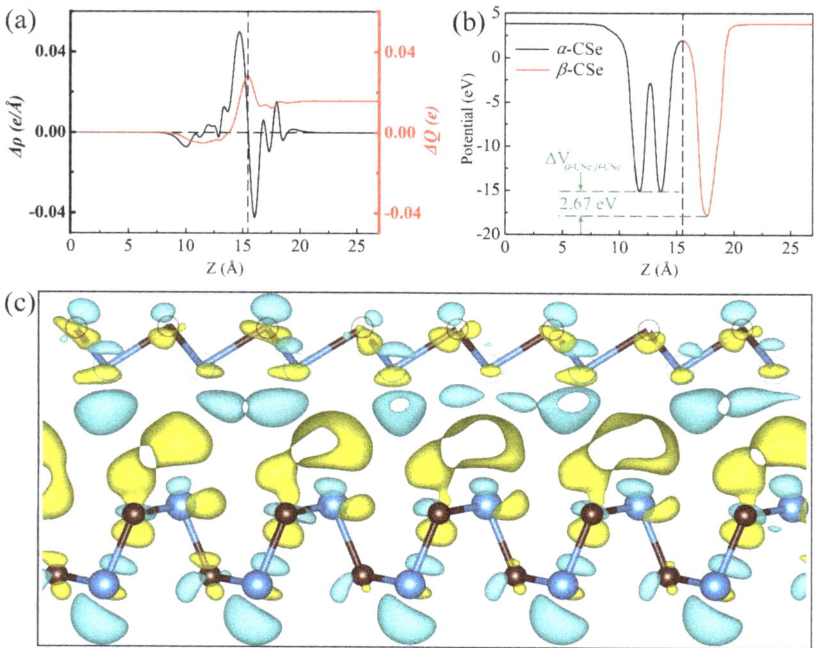

Figure 7. (**a**) The planar-averaged differential charge density $\Delta\rho(z)$ of the α-CSe/β-CSe vdW hetrostructure (black) and the amount of transferred charge $\Delta Q(z)$ as a function of position along the z direction (red). (**b**) xy-averaged electrostatic potential shape through the interface of the α-CSe/β-CSe vdW hetrostructure. (**c**) The sideview of the charge density difference for the α-CSe/β-CSe vdW hetrostructure. The value of isosurfaces is 0.0004 e/Å^3. The yellow and cyan areas exhibit the accumulation and depletion of charges, respectively.

Figure 7b shows the plane-averaged electrostatic potential along the direction normal to the surface of the heterostructure. The potential drop ($\Delta V_{\alpha-CSe/\beta-CSe}$) across the bilayer is found to be 2.67 eV. Such a potential difference indicates an appreciable built-in electronic field across the interface, which may be ascribed to the charge transferred. Meanwhile, the carrier transport is inevitably influenced, i.e., the excitonic behaviors of the α-CSe/β-CSe vdW heterostructure is fairly different from that of the isolated CSe monolayers, because the gradient of the potential across the interface confines the electrons and holes within the different sheets. Bader charge analysis can further support the formation of built-in electronic field. There is about 0.027 e transferring from the β-CSe layer to the α-CSe slab, which is qualitatively consistent with the aforementioned value of $\Delta Q(z)$ at the α-CSe/β-CSe interface. This phenomenon may signify that an apparent space-charge region is formed in this interface. Net positive and negative charges are gathered in different layers, which induce a polarized built-in electric field that was directed from β-CSe to α-CSe. The polarized built-in electric field across the interface imposes another force on carriers (holes and electrons), which is opposite with the diffusion force. The balance between electric field force and diffusion force are propitious to inhibit the recombination of the electrons and holes.

Having inerrably confirmed the type-II heterostructure and clearly understanding a full picture of its charge transfer, it is now imperative to investigate its redox power, as this is responsible for its photocatalytic performance. Erenow, which is the redox power in the isolated sheets, is firstly evaluated. Figure 8 indicates that the band edge alignments before and after contacting and the water redox potentials. The water redox potentials are constant, coming from the experimental measurement [30]. Before contacting, as for α-CSe, the potential of the CBM is positioned at 0.59 eV above the reduction potential of H^+/H_2(−4.44 eV), enabling the generation of H_2, while the oxidation potential of H_2O/O_2 (−5.67 eV) lies below its VBM, indicating that O_2 cannot be spontaneously achieved. The contrary situation takes place in β-CSe, i.e., β-CSe hardly generates H_2 spontaneously, since its CBM is close to the reduction potential of H^+/H_2, while it possesses prominent oxygen evolution ability, since its VBM is largely lower than the oxidation potential of H_2O/O_2. Overall, both the isolated α-CSe and β-CSe monolayers are not suitable as an intrinsic photocatalyst.

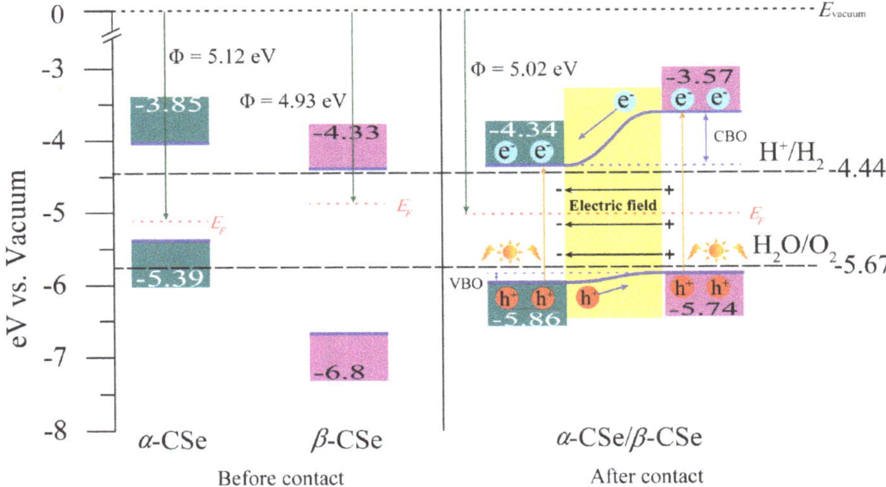

Figure 8. Diagram of the band alignments before and after the isolated monolayer α-CSe and β-CSe contact. The work functions (Φ) for the free-standing monolayer α-CSe, β-CSe and bilayer α-CSe/β-CSe are also provided. The vacuum level E_{vacuum} is set to 0 eV and E_F denotes the Fermi level.

After contacting, the interface charge-transfer drives the Fermi level to move, which brings the movement of band-edge positions of the two CSe sheets. Thus, the band-edges bestride the water redox potentials and make the hybrid α-CSe/β-CSe heterostructure an excellent candidate for applications in sunlight-driven photocatalysis. It is easy to understand the underlying mechanism by the calculated work functions of the free-standing α-CSe and β-CSe monolayers. The work function is defined as $\Phi = E_{vacuum} - E_F$, where, E_{vacuum} and E_F represent the vacuum energy level and Fermi level, respectively. Figure 8 marks the work functions of the free-standing monolayers. A large difference of work function between α-CSe (5.12 eV) and β-CSe (4.93 eV) signifies that the electrons in the β-CSe monolayer will spontaneously transfer to the α-CSe monolayer once they contact each other until the E_F of the two monolayers are aligned. This will result in the heterojunction possessing a nearly middle work function of 5.02 eV in comparison with that of the isolated monolayers. This result matches well with the aforementioned CDD, $\Delta Q(z)$, and Bader charge. Typical type-II band alignment feature also directly resulted in conduction band offset (CBO) and valence band offset (VBO) on both sides of the interface, which is a vital factor in determining the photocatalytic ability Δ_{VBO} and Δ_{CBO}, 0.07 eV and

0.77 eV, enable the photo-generated carriers to participate in the H^+ reduction reaction (hydrogen evolution) and the OH^- oxidation reaction (oxygen evolution).

Actually, as for the water redox reactions, the process has close connection with the built-in interface electric field and the band offset (BO). Under solar light irradiation, the electrons in this heterostructure are excited, where the electrons in valence bands (VBs) thus transfer to conduction bands (CBs), and holes simultaneously leave in their VBs. The built-in electric field across the interface and its BO give rise to the band edges bending. In general, the upward band bending facilitates the holes migrating upward, and it impedes the electrons from moving. Conversely, electrons can transfer downward along the band bending, while holes are not allowed to move freely [59]. More specifically, CBO facilitates the photo-generated electrons to transfer from the CBs of the β-CSe layer to the CBs of the α-CSe layer. On the contrary, VBO promotes the holes to move from the VBs of the α-CSe layer to the VBs of the β-CSe layer. In the meantime, the balance between built-in electronic field and BO prevents the opposite movements of photo-excited carriers. Consequently, under the combined effect of BO and the built-in electric field, those photo-excited carriers are effectively separated and confined in the different layers. Such a separation in space represses the recombination of electron-hole pairs and effectively prolongates their lifetime, which contributes to enhancing the photocatalytic efficiency of the heterojunction. In more detail, massive reductive electrons staying in the CBs of the α-CSe sheet are capable of driving the hydrogen evolution reaction, and simultaneously H_2O/OH^- can also be oxidized to O_2 by substantial oxidizing holes that are located in the VBs of the β-CSe layer. Overall, the α-CSe sheet coupled with the β-CSe monolayer, as a metal-free photocatalyst, enables a higher sunlight-harvest efficiency for photocatalytic water splitting in comparison with the intrinsic unsuitable photocatalytic CSe sheets. Experiments are expected to further corroborate our theoretical findings that were reported in this work [60].

4. Conclusions

In summary, we predicted a stable blue-phosphorus-like monolayer β-CSe that is based on DFT. β-CSe sheet exhibits slightly anisotropic mechanical characteristics: it can endure an ultimate stress of 5.6 N/m at $\varepsilon_{xx} = 0.1$ along the armchair direction, and 5.9 N/m at $\varepsilon_{yy} = 0.14$ along the zigzag direction. As an isoelectronic counterpart of blue phosphorene, β-CSe also presents a wide indirect bandgap (2.37 eV) that is sensitive to the in-plane strain. Thus, the carrier effective mass is strain-dependent. Therefore, a strain-sensitively transport direction displays in β-CSe. The indirect character of band gap is robust in β-CSe, except that armchair-dominant strain can drive an indirect-direct transition. We propose a heterojunction built by the β-CSe sheet covering the α-CSe sheet (α-CSe/β-CSe). The band alignment demonstrates that the α-CSe/β-CSe interface is a type-II van der Waals *p-n* heterojunction. An appreciable built-in electric field across interface, which is caused by the charges transferring from β-CSe layer to α-CSe layer, renders energy bands bending, and causing photo-generated carriers to be spatially well separated. Therefore, α-CSe/β-CSe heterojunction, as a metal-free photocatalyst, is endued an enhanced solar-driven redox ability for photocatalytic water splitting by lessening the electron-hole-pair recombination. This study provides a fundamental insight of designing novel structural phase for high-performance light-emitting devices, and bodes well for application to photocatalysis.

Author Contributions: Q.Z. and Y.F. conceived and designed the research programme; Q.Z. and X.C. performed the calculations; Q.Z., W.Z. and L.W. analyzed the data; Y.F., L.W. and Y.W. polished the English; Q.Z. wrote the paper.

Funding: This research was funded by the National Natural Science Foundation of China under Grant Nos. 11775051.

Conflicts of Interest: The authors declare no conflict of interest.

References

1. Chhowalla, M.; Shin, H.S.; Eda, G.; Li, L.-J.; Loh, K.P.; Zhang, H. The chemistry of two-dimensional layered transition metal dichalcogenide nanosheets. *Nat. Chem.* **2013**, *5*, 263–275. [CrossRef]
2. Li, M.-Y.; Shi, Y.; Cheng, C.-C.; Lu, L.-S.; Lin, Y.-C.; Tang, H.-L.; Tsai, M.-L.; Chu, C.-W.; Wei, K.-H.; He, J.-H.; et al. Epitaxial growth of a monolayer WSe_2-MoS_2 lateral p-n junction with an atomically sharp interface. *Science* **2015**, *349*, 524–528. [CrossRef]
3. Kim, K.S.; Zhao, Y.; Jang, H.; Lee, S.Y.; Kim, J.M.; Kim, K.S.; Ahn, J.-H.; Kim, P.; Choi, J.-Y.; Hong, B.H. Large-scale pattern growth of graphene films for stretchable transparent electrodes. *Nature* **2009**, *457*, 706–710. [CrossRef]
4. Wang, Q.H.; Kalantar-Zadeh, K.; Kis, A.; Coleman, J.N.; Strano, M.S. Electronics and optoelectronics of two-dimensional transition metal dichalcogenides. *Nat. Nanotechnol.* **2012**, *7*, 699–712. [CrossRef] [PubMed]
5. Zhu, Y.; Murali, S.; Stoller, M.D.; Ganesh, K.J.; Cai, W.; Ferreira, P.J.; Pirkle, A.; Wallace, R.M.; Cychosz, K.A.; Thommes, M.; et al. Carbon-Based Supercapacitors Produced by Activation of Graphene. *Science* **2011**, *332*, 1537. [CrossRef]
6. Liu, C.; Yu, Z.; Neff, D.; Zhamu, A.; Jang, B.Z. Graphene-Based Supercapacitor with an Ultrahigh Energy Density. *Nano Lett.* **2010**, *10*, 4863–4868. [CrossRef] [PubMed]
7. Wang, Y.; Wu, Y.; Huang, Y.; Zhang, F.; Yang, X.; Ma, Y.; Chen, Y. Preventing Graphene Sheets from Restacking for High-Capacitance Performance. *J. Phys. Chem. C* **2011**, *115*, 23192–23197. [CrossRef]
8. Ji, L.; Rao, M.; Zheng, H.; Zhang, L.; Li, Y.; Duan, W.; Guo, J.; Cairns, E.J.; Zhang, Y. Graphene Oxide as a Sulfur Immobilizer in High Performance Lithium/Sulfur Cells. *J. Am. Chem. Soc.* **2011**, *133*, 18522–18525. [CrossRef]
9. Hu, L.-H.; Wu, F.-Y.; Lin, C.-T.; Khlobystov, A.N.; Li, L.-J. Graphene-modified $LiFePO_4$ cathode for lithium ion battery beyond theoretical capacity. *Nat. Commun.* **2013**, *4*, 1687. [CrossRef]
10. Xiao, J.; Mei, D.; Li, X.; Xu, W.; Wang, D.; Graff, G.L.; Bennett, W.D.; Nie, Z.; Saraf, L.V.; Aksay, I.A.; et al. Hierarchically Porous Graphene as a Lithium–Air Battery Electrode. *Nano Lett.* **2011**, *11*, 5071–5078. [CrossRef]
11. Huang, L.; Li, Y.; Wei, Z.; Li, J. Strain induced piezoelectric effect in black phosphorus and MoS_2 van der Waals heterostructure. *Sci. Rep.* **2015**, *5*, 16448. [CrossRef] [PubMed]
12. Liu, J.; Liu, Y.; Liu, N.; Han, Y.; Zhang, X.; Huang, H.; Lifshitz, Y.; Lee, S.-T.; Zhong, J.; Kang, Z. Metal-free efficient photocatalyst for stable visible water splitting via a two-electron pathway. *Science* **2015**, *347*, 970–974. [CrossRef]
13. Shalom, M.; Gimenez, S.; Schipper, F.; Herraiz-Cardona, I.; Bisquert, J.; Antonietti, M. Controlled Carbon Nitride Growth on Surfaces for Hydrogen Evolution Electrodes. *Angew. Chem.* **2014**, *126*, 3728–3732. [CrossRef]
14. Chen, Y.; Sun, H.; Peng, W. 2D Transition Metal Dichalcogenides and Graphene-Based Ternary Composites for Photocatalytic Hydrogen Evolution and Pollutants Degradation. *Nanomaterials* **2017**, *7*, 62. [CrossRef]
15. Li, L.; Yu, Y.; Ye, G.J.; Ge, Q.; Ou, X.; Wu, H.; Feng, D.; Chen, X.H.; Zhang, Y. Black phosphorus field-effect transistors. *Nat. Nanotechnol.* **2014**, *9*, 372–377. [CrossRef]
16. Nazir, G.; Khan, F.M.; Aftab, S.; Afzal, M.A.; Dastgeer, G.; Rehman, A.M.; Seo, Y.; Eom, J. Gate Tunable Transport in Graphene/MoS_2/(Cr/Au) Vertical Field-Effect Transistors. *Nanomaterials* **2018**, *8*, 14. [CrossRef] [PubMed]
17. Kou, L.; Frauenheim, T.; Chen, C. Phosphorene as a Superior Gas Sensor: Selective Adsorption and Distinct I-V Response. *J. Phys. Chem. Lett.* **2014**, *5*, 2675–2681. [CrossRef] [PubMed]
18. Lee, C.; Wei, X.; Kysar, J.W.; Hone, J. Measurement of the Elastic Properties and Intrinsic Strength of Monolayer Graphene. *Science* **2008**, *321*, 385–388. [CrossRef]
19. Geim, A.K.; Novoselov, K.S. The rise of graphene. *Nat. Mater.* **2007**, *6*, 183–191. [CrossRef]
20. Castro Neto, A.H.; Guinea, F.; Peres, N.M.R.; Novoselov, K.S.; Geim, A.K. The electronic properties of graphene. *Rev. Mod. Phys.* **2009**, *81*, 109–162. [CrossRef]
21. Mak, K.F.; Lee, C.; Hone, J.; Shan, J.; Heinz, T.F. Atomically Thin MoS_2: A New Direct-Gap Semiconductor. *Phys. Rev. Lett.* **2010**, *105*, 136805. [CrossRef]
22. Radisavljevic, B.; Radenovic, A.; Brivio, J.; Giacometti, V.; Kis, A. Single-layer MoS_2 transistors. *Nat. Nanotechnol.* **2011**, *6*, 147–150. [CrossRef]

23. Lu, J.; Carvalho, A.; Chan, X.K.; Liu, H.; Liu, B.; Tok, E.S.; Loh, K.P.; Castro Neto, A.H.; Sow, C.H. Atomic Healing of Defects in Transition Metal Dichalcogenides. *Nano Lett.* **2015**, *15*, 3524–3532. [CrossRef]
24. Sa, B.; Li, Y.-L.; Qi, J.; Ahuja, R.; Sun, Z. Strain Engineering for Phosphorene: The Potential Application as a Photocatalyst. *J. Phys. Chem. C* **2014**, *118*, 26560–26568. [CrossRef]
25. Qiao, J.; Kong, X.; Hu, Z.-X.; Yang, F.; Ji, W. High-mobility transport anisotropy and linear dichroism in few-layer black phosphorus. *Nat. Commun.* **2014**, *5*, 4475. [CrossRef]
26. Yang, W.; Chen, G.; Shi, Z.; Liu, C.-C.; Zhang, L.; Xie, G.; Cheng, M.; Wang, D.; Yang, R.; Shi, D.; et al. Epitaxial growth of single-domain graphene on hexagonal boron nitride. *Nat. Mater.* **2013**, *12*, 792–797. [CrossRef]
27. Cai, Y.; Zhang, G.; Zhang, Y.-W. Electronic Properties of Phosphorene/Graphene and Phosphorene/Hexagonal Boron Nitride Heterostructures. *J. Phys. Chem. C* **2015**, *119*, 13929–13936. [CrossRef]
28. Huang, L.; Wu, F.; Li, J. Structural anisotropy results in strain-tunable electronic and optical properties in monolayer GeX and SnX (X = S, Se, Te). *J. Chem. Phys.* **2016**, *144*, 114708. [CrossRef]
29. Kamal, C.; Chakrabarti, A.; Ezawa, M. Direct band gaps in group IV-VI monolayer materials: Binary counterparts of phosphorene. *Phys. Rev. B* **2016**, *93*, 125428. [CrossRef]
30. Zhang, Q.; Chen, X.; Liu, W.-C.; Wang, Y. Stacking effect on electronic, photocatalytic and optical properties: A comparison between bilayer and monolayer SnS. *Comput. Mater. Sci.* **2019**, *158*, 272–281. [CrossRef]
31. Zhang, Q.; Xin, T.; Lu, X.; Wang, Y. Optoelectronic Properties of X-Doped (X = O, S, Te) Photovoltaic CSe with Puckered Structure. *Materials* **2018**, *11*, 431. [CrossRef]
32. Brent, J.R.; Lewis, D.J.; Lorenz, T.; Lewis, E.A.; Savjani, N.; Haigh, S.J.; Seifert, G.; Derby, B.; O'Brien, P. Tin(II) Sulfide (SnS) Nanosheets by Liquid-Phase Exfoliation of Herzenbergite: IV-VI Main Group Two-Dimensional Atomic Crystals. *J. Am. Chem. Soc.* **2015**, *137*, 12689–12696. [CrossRef]
33. Jiang, J.; Wong, C.P.Y.; Zou, J.; Li, S.; Wang, Q.; Chen, J.; Qi, D.; Wang, H.; Eda, G.; Chua, D.H.C.; et al. Two-step fabrication of single-layer rectangular SnSe flakes. *2D Mater.* **2017**, *4*, 021026. [CrossRef]
34. Sirikumara, H.I.; Jayasekera, T. Tunable indirect-direct transition of few-layer SnSe via interface engineering. *J. Phys. Condens. Matter.* **2017**, *29*, 425501. [CrossRef]
35. Zhu, Z.; Tomanek, D. Semiconducting layered blue phosphorus: A computational study. *Phys. Rev. Lett.* **2014**, *112*, 176802. [CrossRef]
36. Zeng, J.; Cui, P.; Zhang, Z. Half Layer By Half Layer Growth of a Blue Phosphorene Monolayer on a GaN(001) Substrate. *Phys. Rev. Lett.* **2017**, *118*, 046101. [CrossRef]
37. Zhang, J.L.; Zhao, S.; Han, C.; Wang, Z.; Zhong, S.; Sun, S.; Guo, R.; Zhou, X.; Gu, C.D.; Yuan, K.D.; et al. Epitaxial Growth of Single Layer Blue Phosphorus: A New Phase of Two-Dimensional Phosphorus. *Nano Lett.* **2016**, *16*, 4903–4908. [CrossRef]
38. Zhu, Z.; Guan, J.; Liu, D.; Tománek, D. Designing Isoelectronic Counterparts to Layered Group V Semiconductors. *ACS Nano* **2015**, *9*, 8284–8290. [CrossRef]
39. Hu, Z.Y.; Li, K.Y.; Lu, Y.; Huang, Y.; Shao, X.H. High thermoelectric performances of monolayer SnSe allotropes. *Nanoscale* **2017**, *9*, 16093–16100. [CrossRef]
40. Huang, L.; Li, J. Tunable electronic structure of black phosphorus/blue phosphorus van der Waals p-n heterostructure. *Appl. Phys. Lett.* **2016**, *108*, 083101. [CrossRef]
41. Kresse, G.; Hafner, J. Ab initio molecular dynamics for liquid metals. *Phys. Rev. B* **1993**, *47*, 558–561. [CrossRef]
42. Hafner, J. Ab-initio simulations of materials using VASP: Density-functional theory and beyond. *J. Comput. Chem.* **2008**, *29*, 2044–2078. [CrossRef]
43. Perdew, J.P.; Wang, Y. Accurate and simple analytic representation of the electron-gas correlation energy. *Phys. Rev. B* **1992**, *45*, 13244. [CrossRef]
44. Perdew, J.P.; Burke, K.; Ernzerhof, M. Generalized gradient approximation made simple (vol 77, pg 3865, 1996). *Phys. Rev. Lett.* **1997**, *78*, 1396. [CrossRef]
45. Blöchl, P.E. Projector augmented-wave method. *Phys. Rev. B* **1994**, *50*, 17953. [CrossRef]
46. Mi, K.; Xie, J.; Si, M.S.; Gao, C.X. Layer-stacking effect on electronic structures of bilayer arsenene. *EPL* **2017**, *117*, 27002. [CrossRef]
47. Heyd, J.; Scuseria, G.E.; Ernzerhof, M. Hybrid functionals based on a screened Coulomb potential. *J. Chem. Phys.* **2003**, *118*, 8207–8215. [CrossRef]

48. Togo, A.; Oba, F.; Tanaka, I. First-principles calculations of the ferroelastic transition between rutile-type and$CaCl_2$-typeSiO_2 at high pressures. *Phys. Rev. B* **2008**, *78*, 134106. [CrossRef]
49. Shi, L.-B.; Zhang, Y.-Y.; Xiu, X.-M.; Dong, H.-K. Structural characteristics and strain behavior of two-dimensional C_3N: First principles calculations. *Carbon* **2018**, *134*, 103–111. [CrossRef]
50. Wang, H.; Li, Q.; Gao, Y.; Miao, F.; Zhou, X.-F.; Wan, X.G. Strain effects on borophene: Ideal strength, negative Possion's ratio and phonon instability. *New J. Phys.* **2016**, *18*, 073016. [CrossRef]
51. Gao, Y.; Zhang, L.; Yao, G.; Wang, H. Unique mechanical responses of layered phosphorus-like group-IV monochalcogenides. *J. Appl. Phys.* **2019**, *125*, 082519. [CrossRef]
52. Wang, H.; Liu, E.; Wang, Y.; Wan, B.; Ho, C.-H.; Miao, F.; Wan, X.G. Cleavage tendency of anisotropic two-dimensional materials: ReX_2 (X = S, Se) and WTe_2. *Phys. Rev. B* **2017**, *96*, 165418. [CrossRef]
53. Jiang, J.W.; Park, H.S. Negative poisson's ratio in single-layer black phosphorus. *Nat. Commun.* **2014**, *5*, 4727. [CrossRef]
54. Liu, F.; Ming, P.; Li, J. Ab initio calculation of ideal strength and phonon instability of graphene under tension. *Phys. Rev. B* **2007**, *76*, 064120. [CrossRef]
55. Yang, C.; Yu, Z.; Lu, P.; Liu, Y.; Ye, H.; Gao, T. Phonon instability and ideal strength of silicene under tension. *Comput. Mater. Sci.* **2014**, *95*, 420–428. [CrossRef]
56. Huang, M.; Yan, H.; Heinz, T.F.; Hone, J. Probing Strain-Induced Electronic Structure Change in Graphene by Raman Spectroscopy. *Nano Lett.* **2010**, *10*, 4074–4079. [CrossRef]
57. Peng, X.; Wei, Q.; Copple, A. Strain-engineered direct-indirect band gap transition and its mechanism in two-dimensional phosphorene. *Phys. Rev. B* **2014**, *90*, 085402. [CrossRef]
58. Ghosh, B.; Nahas, S.; Bhowmick, S.; Agarwal, A. Electric field induced gap modification in ultrathin blue phosphorus. *Phys. Rev. B* **2015**, *91*, 115433. [CrossRef]
59. Wang, J.; Li, X.; You, Y.; Yang, X.; Wang, Y.; Li, Q. Interfacial coupling induced direct Z-scheme water splitting in metal-free photocatalyst: C_3N/g-C_3N_4 heterojunctions. *Nanotechnology* **2018**, *29*, 365401. [CrossRef]
60. Martín-Sánchez, J.; Trotta, R.; Piredda, G.; Schimpf, C.; Trevisi, G.; Seravalli, L.; Frigeri, P.; Stroj, S.; Lettner, T.; Reindl, M.; et al. Reversible Control of In-Plane Elastic Stress Tensor in Nanomembranes. *Adv. Opt. Mater.* **2016**, *4*, 682–687. [CrossRef]

Article

Physical Properties and Photovoltaic Application of Semiconducting Pd_2Se_3 Monolayer

Xiaoyin Li [1,2], Shunhong Zhang [3], Yaguang Guo [1,2], Fancy Qian Wang [1,2] and Qian Wang [1,2,*]

1 Center for Applied Physics and Technology, HEDPS, Department of Materials Science and Engineering, College of Engineering, Peking University, Beijing 100871, China; lixiaoyin@pku.edu.cn (X.L.); guoyaguang@pku.edu.cn (Y.G.); qianwang7@pku.edu.cn (F.Q.W.)
2 Collaborative Innovation Center of IFSA (CICIFSA), Shanghai Jiao Tong University, Shanghai 200240, China
3 Institute for Advanced Study, Tsinghua University, Beijing 100084, China; zhangshunhong@tsinghua.edu.cn
* Correspondence: qianwang2@pku.edu.cn; Tel.: +86-10-62755644

Received: 18 September 2018; Accepted: 13 October 2018; Published: 14 October 2018

Abstract: Palladium selenides have attracted considerable attention because of their intriguing properties and wide applications. Motivated by the successful synthesis of Pd_2Se_3 monolayer (Lin et al., Phys. Rev. Lett., 2017, 119, 016101), here we systematically study its physical properties and device applications using state-of-the-art first principles calculations. We demonstrate that the Pd_2Se_3 monolayer has a desirable quasi-direct band gap (1.39 eV) for light absorption, a high electron mobility (140.4 $cm^2V^{-1}s^{-1}$) and strong optical absorption (~10^5 cm^{-1}) in the visible solar spectrum, showing a great potential for absorber material in ultrathin photovoltaic devices. Furthermore, its bandgap can be tuned by applying biaxial strain, changing from indirect to direct. Equally important, replacing Se with S results in a stable Pd_2S_3 monolayer that can form a type-II heterostructure with the Pd_2Se_3 monolayer by vertically stacking them together. The power conversion efficiency (PCE) of the heterostructure-based solar cell reaches 20%, higher than that of $MoS_2/MoSe_2$ solar cell. Our study would motivate experimental efforts in achieving Pd_2Se_3 monolayer-based heterostructures for new efficient photovoltaic devices.

Keywords: palladium selenide monolayer; physical properties; light-harvesting performance; type-II heterostructure; first principles calculations

1. Introduction

Two-dimensional (2D) transition metal chalcogenides (TMCs), including semiconducting MoS_2 [1], $MoSe_2$ [2], WS_2 [3], WSe_2 [4], ReS_2 [5], PtS_2 [6], $PdSe_2$ [7,8], and metallic VS_2 [9] and NbS_2 [10] are of current interest because of their extraordinary properties and practical applications in catalysis [11], electronics [12–14], optoelectronics [15,16] and valleytronics [17,18]. Among them, the layered $PdSe_2$ has attracted special attention due to its unique atomic configuration and electronic properties [8,19,20]. Whereas previous studies mainly focused on the $PdSe_2$ monolayer that has the same structural form as a single layer of the bulk $PdSe_2$ [7,21,22]. Very recently, Lin et al. reported the successful exfoliation of a new monolayer phase with a stoichiometry of Pd_2Se_3 [23], and found that Se vacancies in the pristine $PdSe_2$ reduce the distance between the layers, melding the two layers into one, thus, resulting in the formation of the Pd_2Se_3 monolayer. Due to its structural novelty, subsequent efforts have been made to further explore this new material, including its electronic and optical properties [24] and thermoelectric performance [25], as well as theoretical calculations and experimental synthesis of the lateral junctions between a $PdSe_2$ bilayer and the Pd_2Se_3 monolayer [26].

We noticed that in Reference [24] the results were obtained from standard density functional theory (DFT) calculations (the Perdew-Burke-Ernzerhof (PBE) functional [27] for the generalized gradient approximation (GGA)), which is well-known to underestimate the electronic band gap of semiconductors. However, the accurate description of electronic structure is important for further investigation of electronic and optical properties. To overcome this limitation, various theoretical approaches have been developed. Among them, the hybrid functional that combines standard DFT with Hartree-Fock (HF) calculations has been widely used for calculating the band gaps, because it predicts more reliable physical properties and keeps a good compromise with computational efficiency. Therefore, we use the Heyd-Scuseria-Ernzerhof (HSE06) hybrid functional [28,29] to study the electronic, transport and optical properties of the newly synthesized Pd_2Se_3 monolayer. We show that this monolayer possesses a desirable bandgap for light harvesting, offering better opportunity for photovoltaic applications. Moreover, its electronic structure can be effectively tuned by applying biaxial strain, and indirect to direct bandgap transition occurs with a small critical strain of 2%. In addition, a stable Pd_2S_3 monolayer can be formed by substituting Se with S, which can be used to construct a type-II heterostructure with the Pd_2Se_3 monolayer. The heterostructure-based solar cell can reach a high power conversion efficiency (PCE) of 20%. These fascinating properties make the Pd_2Se_3 monolayer a promising candidate for future applications in nanoscale electronics and photonics.

2. Computational Methods

Within the framework of DFT, our first-principles calculations are performed using the projector augmented wave (PAW) method [30] as implemented in the Vienna Ab initio Simulation Package (VASP) [31]. The Perdew-Burke-Ernzerhof (PBE) functional [27] with the generalized gradient approximation (GGA) is used to treat the electron exchange-correlation interactions in crystal structure calculations, while the Heyd-Scuseria-Ernzerhof (HSE06) hybrid functional [28,29], which includes the Hartree-Fock exchange energy and the Coulomb screening effect, is used to calculate the electronic and optical properties. A kinetic energy cutoff of 350 eV is set for the plane wave basis. The convergence criteria are 10^{-5} eV and 10^{-3} eV/Å for total energy and atomic force components, respectively. The Brillouin zone is represented by k points with a grid density of $2\pi \times 0.02$ Å^{-1} in the reciprocal space using the Monkhorst-Pack scheme [32]. An adequate vacuum space (~20 Å) in the direction perpendicular to the sheet is used to minimize the interlayer interactions under the periodic boundary condition. Spin-orbit coupling (SOC) interactions are not included since our calculation shows that the SOC has negligible effect on electronic structure of the monolayer (see Figure S2). Phonon dispersion and density of states (DOS) are calculated using the finite displacement method [33] as implemented in the Phonopy code [34].

In the calculation of carrier mobility (μ), we consider the perfect crystal of the monolayer without defects and impurities. In addition, carrier mobility is a function of temperature. We set the temperature to be 300 K in our calculation, since most devices work at room temperature. In this situation, the dominant source of electron scattering is from acoustic phonons and the carrier mobility can be obtained using deformation potential theory proposed by Bardeen and Shockley [35], which has been successfully employed in many 2D materials [36–39]. Using effective mass approximation, the analytical expression of carrier mobility in 2D materials can be written as

$$\mu = \frac{e\hbar^3 C}{k_B T m^* m_d E_1^2} \tag{1}$$

C is the elastic modulus of the 2D sheet, T is the temperature, which is taken to be 300 K in our calculations, $m^* = \hbar^2[\partial^2 E(k)/\partial k^2]^{-1}$ is the effective mass of the band edge carrier along the transport direction and m_d is the average effective mass determined by $m_d = \sqrt{m_x^* m_y^*}$. E_1 is the DP constant defined as the energy shift of the band edge with respect to lattice dilation and compression, and k_B and $\hbar$ are Boltzmann and reduced Planck constants, respectively.

The optical absorption coefficient (α) can be expressed as [40–42]

$$\alpha(\omega) = \sqrt{2}\omega\left[\sqrt{\varepsilon_1^2(\omega) + \varepsilon_2^2(\omega)} - \varepsilon_1(\omega)\right]^{1/2} \tag{2}$$

where $\varepsilon_1(\omega)$ and $\varepsilon_2(\omega)$ are the real and imaginary parts of the frequency-dependent dielectric functions which are obtained using the time-dependent Hartree-Fock approach (TDHF) based on the HSE06 hybrid functional calculations [43]. The model, developed by Scharber et al. for organic solar cells [44] and exciton-based 2D solar cells [45–49], is used to calculate the maximum PCE in the limit of 100% external quantum efficiency (EQE), which can be written as

$$\eta = \frac{\beta_{FF} V_{oc} J_{sc}}{P_{solar}} = \frac{0.65(E_g^d - \Delta E_c - 0.3)\int_{E_g^d}^{\infty} \frac{P(\hbar\omega)}{\hbar\omega} d(\hbar\omega)}{\int_0^{\infty} P(\hbar\omega) d(\hbar\omega)} \quad (3)$$

Here, the fill factor β_{FF}, which is the ratio of maximum power output to the product of the open-circuit voltage (V_{oc}) and the short-circuit current (J_{sc}), is estimated to be 0.65 in this model. V_{oc} (in eV) is estimated by the term ($E_g^d - \Delta E_c - 0.3$), where E_g^d is the bandgap of the donor and ΔE_c is the conduction band (CB) offset between donor and acceptor. J_{sc} is obtained by $\int_{E_g^d}^{\infty} \frac{P(\hbar\omega)}{\hbar\omega} d(\hbar\omega)$, and the total incident solar power per unit area P_{solar} is equal to $\int_0^{\infty} P(\hbar\omega) d(\hbar\omega)$. Here, $\hbar$ and ω are reduced Planck constants and photon frequency, and $P(\hbar\omega)$ is the air mass (AM) 1.5 solar energy flux (expressed in W m^{-2} eV^{-1}) at the photon energy ($\hbar\omega$).

3. Results and Discussion

3.1. Geometric Structure of Pd_2Se_3

Figure 1a,b shows the optimized monolayer structures of Pd_2Se_3 and $PdSe_2$ respectively (the structural details are listed in Table S1, Supplementary Materials). For simplicity, we refer these two monolayer structures as Pd_2Se_3 and $PdSe_2$ in the following discussions unless stated otherwise. There are similarities as well as differences between the two structures. On one hand, they both consist of a layer of metal Pd atoms sandwiched between the two layers of chalcogen Se atoms, and each Pd atom binds to four Se atoms forming the square-planar ($PdSe_4$) structural units. On the other hand, Pd_2Se_3 indeed distinguishes itself from $PdSe_2$ in the following characteristics: (1) Pd_2Se_3 possesses *Pmmn* symmetry (point group D_{2h}) with four Pd and six Se atoms in one unit cell. While the symmetry of $PdSe_2$ is $P2_1/c$ (point group C_{2h}) and each unit cell contains two Pd and four Se atoms. The different crystal symmetries result in distinct resonance in the Raman spectroscopy, which can serve as an efficient and straightforward clue for experimentalists to confirm the formation of Pd_2Se_3. The details about the calculated Raman spectra of Pd_2Se_3 and $PdSe_2$ are presented in the Supporting Information (Figure S1). (2) There are two chemically nonequivalent Se in Pd_2Se_3, marked in orange (Se2) and yellow (Se1) respectively. The two neighboring Se2 atoms form a covalent Se-Se bond while each Se1 atom is unpaired and binds to four neighboring Pd atoms. Whereas in $PdSe_2$, all Se atoms are in dimers and form the Se-Se bonds. (3) In Pd_2Se_3, the Se-Se dumbbells are parallel to the Pd layer, while in $PdSe_2$, they cross the Pd layer. (4) Pd_2Se_3 and $PdSe_2$ have different charge-balanced formulas, written as $(Pd^{2+})_2(Se^{2-})(Se_2{}^{2-})$ and $(Pd^{2+})(Se_2{}^{2-})$ respectively, due to the different chemical environments of Se atoms in the two structures. Since the properties of materials are essentially determined by their geometric structures, one can expect that Pd_2Se_3 would possess some new and different properties from those of $PdSe_2$.

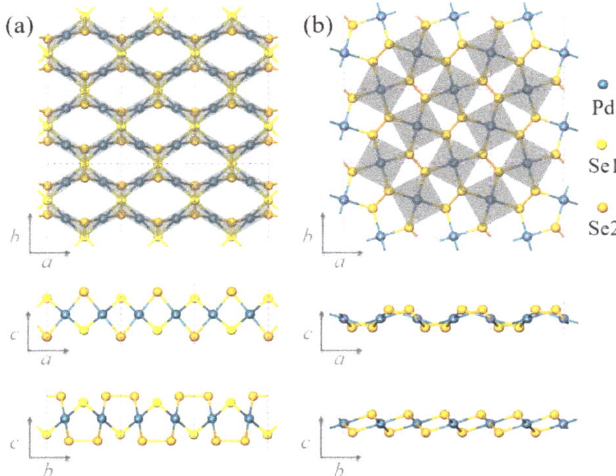

Figure 1. Optimized atomic structure of (**a**) Pd_2Se_3, and (**b**) $PdSe_2$ monolayers. The gray tetragons and purple dashed rectangles correspond to the planar ($PdSe_4$) units and the primitive cells of the two structures, respectively.

3.2. Electronic Properties of Pd_2Se_3

We then investigated the electronic properties of Pd_2Se_3 by calculating its electronic band structure and density of states (DOS) using the hybrid HSE06 functional. Figure 2a shows the calculated band structure around the Fermi level and corresponding total and partial DOS. The bandgap size of Pd_2Se_3 is 1.39 eV, close to the optimum value (~1.3 eV) for solar cell materials [50–52]. Although Pd_2Se_3 is an indirect bandgap semiconductor with the valence band maximum (VBM) located at the Γ point and the conduction band minimum (CBM) located on the Y-M path, Pd_2Se_3 can be considered as a quasi-direct bandgap semiconductor because of the existence of the sub-CBM at the Γ point (CB2) that is only marginally higher in energy than the true CBM (the energy difference is less than 50 meV). The weakly indirect bandgap is desirable for photovoltaic applications since it can simultaneously increase optical absorbance and photocarrier lifetimes [41,53,54]. To assess the effect of SOC interaction, we computed the band structure of Pd_2Se_3 at the level of HSE06+SOC. The results in Figure S2 reveal that the SOC in Pd_2Se_3 is weak and has negligible effect on the bandgap of this structure. Hereafter, we do not include the SOC interaction and just use the HSE06 scheme for calculations in this study.

An analysis of the partial DOS in Figure 2a indicates that the electronic states of valance and conduction bands mainly originate from Se 4*p* and Pd 4*d* orbitals. In addition, the overlap of the orbital-projected DOS implies strong hybridization, that is, the formation of covalent bonds between Se 4*p* and Pd 4*d* orbitals. By calculating wave functions for the VBM and CBM, we visualized their electronic states showing distinct antibonding features for both of them (see Figure 2b). However, to gain a better understanding for the covalent bonding in this 2D structure, the electronic bands not only limited to near the Fermi level but also in a large energy range should be taken into account.

Figure 2c displays the band structure and partial DOS including all occupied and sufficient unoccupied states of Pd_2Se_3. Combining crystal field theory and crystal structure chemistry analysis, we can clearly identify the electronic states in the energy range from -17 to 4 eV. From partial DOS, the bands in the energy range of $-17 \sim -12.5$ eV are primarily from Se 4*s* orbitals. According to the different bonding states, they can be classified into three groups. The bottom and upper subsets correspond to the bonding and antibonding states dominated by the formation of Se-Se bonds, and the middle part is the nonbonding state of the unpaired Se 4*s* orbitals. When the energy goes up, there occurs Pd 4*d* orbitals. In Pd_2Se_3, the Pd atom is coordinated in a nearly perfect square-planar geometry, and its $4d$ orbitals split into four energy levels, i.e., e_g (d_{xz}/d_{yz}), a_{1g} (d_{z2}), b_{2g} (d_{xy}), and b_{1g} ($d_{x2\text{-}y2}$) from low to high energy. These d orbitals overlapping with Se $4p$ orbitals constitutes the bands in the energy range of -7.5 ~4 eV. In Figure 2d, we present the schematic drawing of DOS and energy level diagram of Pd_2Se_3 to explain details about how Pd $4d$ orbitals interact with Se $4p$ orbitals. It shows that a_{1g}, b_{2g}, and b_{1g} orbitals hybridize with Se $4p$ orbitals leading to lower energy bonding states and higher energy antibonding states, whereas the nonbonding states in the energy range from -4 to -1.7 eV stemming mainly from e_g orbitals. More importantly, the bandgap that separates occupied and unoccupied states lies in the antibonding region and amounts to the splitting energy between and states, consistent with the results of wave functions for the VBM and CBM in Figure 2b. The systematic and deep exploration of electronic structure of Pd_2Se_3 is crucial for understanding its properties and origins of intriguing physical phenomena.

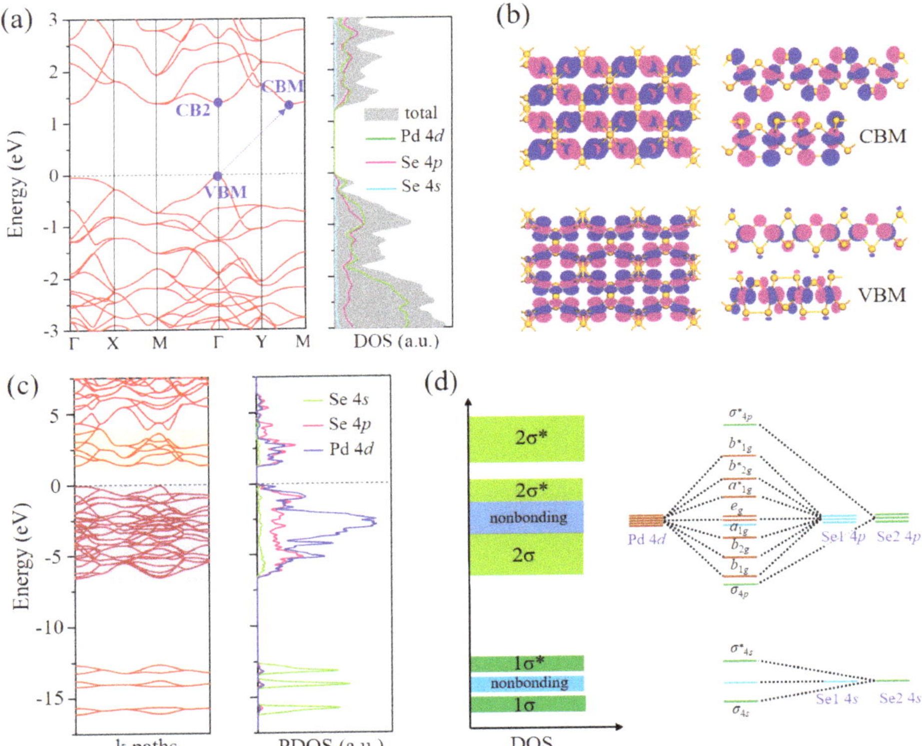

Figure 2. (**a**) Band structure and DOS around the Fermi level. The VBM and CB(M) are marked by blue dots; (**b**) Spatial visualization of wave functions for the VBM and CBM, using an isosurface of 0.04 eÅ^{-3}; (**c**) Band structure and partial DOS with all valence states included. (**d**) Schematics of DOS and energy level diagram.

3.3. Strain Engineering of Electronic Band Structure of Pd_2Se_3

From above electronic structure analysis, it is clear that Pd_2Se_3 is a covalent semiconductor, and a connection between elastic strain and its electronic structure is expected. This is because elastic strain generally weakens the covalent interaction as the bonds lengthen, exerting efficient modulation on the band energies and bandgap. For this reason, we applied a biaxial tensile strain to Pd_2Se_3 and study its effect on the electronic bands of Pd_2Se_3.

Figure 3a shows the evolution of band structure with biaxial strain varying from 0% to 9%. It indicates that both direct and indirect bandgaps increase, and a transition from indirect bandgap to direct bandgap occurs when the biaxial strain is applied. To acquire a more accurate energy profile, we present the strain-dependent bandgaps in Figure 3b, which clearly shows the increasing trend of bandgaps and the bandgap transition from indirect to direct at the critical strain of 2%.

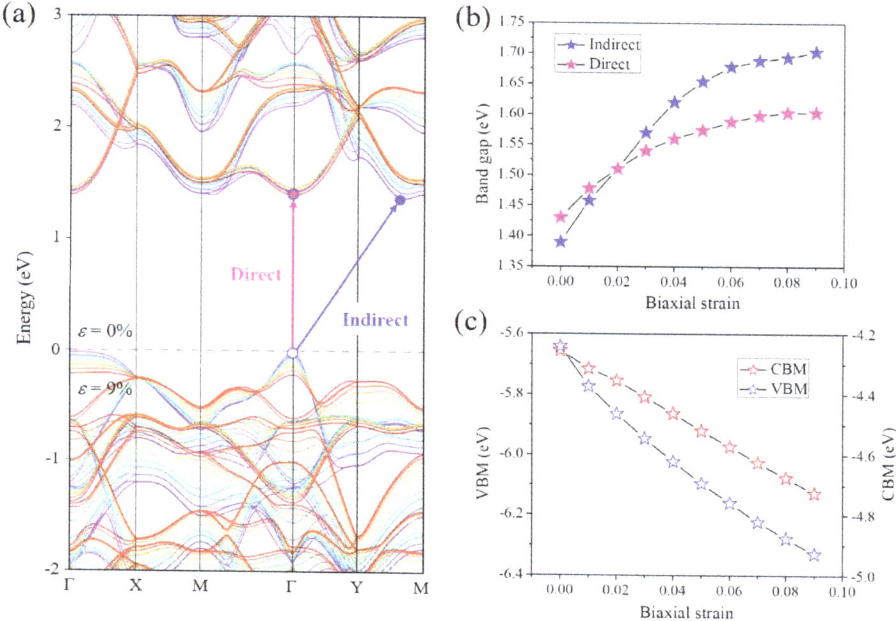

Figure 3. (**a**) Electronic band structure of the Pd_2Se_3 monolayer under biaxial strains varying from 0% (violet line) to 9% (red line); (**b**) Direct and indirect bandgaps under different biaxial strains; (**c**) Biaxial strain-dependent energies of the VBM and CBM with respect to the vacuum level. All calculations are based on the HSE06 functional.

The strain-dependent bandgap of Pd_2Se_3 can be understood by analyzing its electronic structure in Figure 2, which shows that the valence and conduction bands both originate from antibonding states. Application of a tensile strain increases the Pd–Se bond length thus decreases the amount of orbital overlap, leading to the stabilization of valence and conduction bands and reducing them in energy. This is consistent with our results, as shown in Figure 3c, which displays the strain-dependent energy levels for the VBM and CBM. However, since the biaxial tensile strain not only enlarges the bond length but also distorts the square-planar ligand field (see Figure S3 for details), the valence band responds more strongly to strains than the conduction band, resulting in the increase of bandgap in the imposed strain filed.

Additionally, we also examined the structural stability under biaxial strains. The phonon dispersion in Figure S4 demonstrates that the structure remains stable under the strain of 9%. The large strain tolerance and an electronic structure that has a continuous response in the imposed strain field indicate the great potential of Pd_2Se_3 in future flexible electronics.

3.4. Transport Properties of Pd_2Se_3

We also investigated the transport properties of Pd_2Se_3 by calculating its room-temperature carrier mobilities as summarized in Table 1. One can see that the mobilities for both electrons and holes are slightly anisotropic along the x and y directions, due to the structural anisotropy of Pd_2Se_3. Meanwhile, the electron mobility along the y direction is estimated to be 140.4 $cm^2V^{-1}s^{-1}$, significantly higher than that of hole. When compared with $PdSe_2$, whose carrier mobilities are calculated at the same theoretical level and listed in Table 1, Pd_2Se_3 possesses higher electron mobility and lower hole mobility, showing strong asymmetry in electron and hole transport. Although the carrier mobilities of Pd_2Se_3 is lower than the theoretical predicted carrier mobilities of some other 2D materials [36,38,55], it is still commendable if realized in practice [20].

Table 1. Calculated deformation potential constant (E_1), elastic modulus (C), effective mass (m^*), and mobility (μ) for electron and hole in the x and y directions for Pd_2Se_3 and $PdSe_2$ monolayers at 300 K.

	Carrier Type	E_1 (eV)	C (N/m)	m^* (m_e)	μ ($cm^2V^{-1}s^{-1}$)
Pd_2Se_3	electron (x)	3.756	33.02	0.762	101.9
	electron (y)	3.785	32.93	0.543	140.4
	hole (x)	2.870	33.02	9.029	7.3
	hole (y)	12.082	32.93	0.187	19.9
$PdSe_2$	electron (x)	9.542	32.45	0.429	43.36
	electron (y)	9.982	55.05	0.390	73.99
	hole (x)	3.352	32.45	0.656	97.94
	hole (y)	3.074	55.05	1.401	92.54

3.5. Optical Properties of Pd_2Se_3

Attracted by the suitable bandgap and intriguing electronic properties of Pd_2Se_3, we further explored its light-harvesting performance by calculating the dielectric functions based on the hybrid HSE06 functional. Figure 4a shows the real (ε_1) and imaginary (ε_2) parts of the frequency-dependent complex dielectric functions of Pd_2Se_3. With the dielectric functions, we derive its optical absorption coefficient (α), as shown in Figure 4b. For comparison, the absorption spectra of $PdSe_2$ was also calculated. We notice that, for both Pd_2Se_3 and $PdSe_2$, the overall absorption coefficients are close to the order of 10^5 cm^{-1} and only show little difference along the x and y directions, which are considerably desirable for optical absorption. Moreover, as shown in Figure 4b, the absorption coefficient of Pd_2Se_3 is slightly larger than that of $PdSe_2$ in nearly the entire of the energy range, indicating the improved light-harvesting performance of Pd_2Se_3 as compared with $PdSe_2$. Furthermore, we also investigated the biaxial strain influence on the optical performance of Pd_2Se_3. The calculated strain-dependent optical absorption spectra are presented in Figure 4c. It shows that the strain slightly affects the optical absorption of Pd_2Se_3, which is favorable for applications in flexible systems since it guarantees steady performance of devices under stretching.

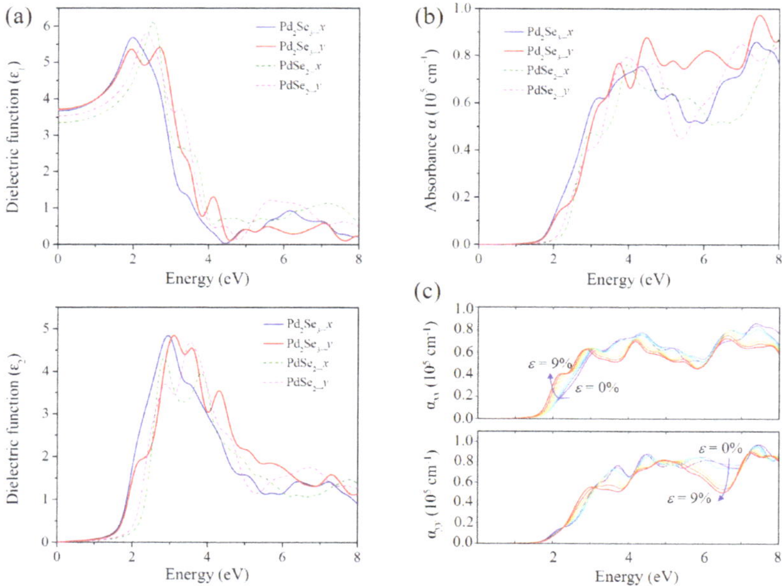

Figure 4. (**a**) Real part (ε_1) and imaginary part (ε_2) of the complex dielectric function, and (**b**) optical absorption spectra of Pd_2Se_3, as compared to those of $PdSe_2$ along the x and y directions respectively; (**c**) Optical absorption spectra of Pd_2Se_3 under different biaxial strains from 0% (violet line) to 9% (red line).

3.6. Extension and Photovoltaic Application of Pd_2Se_3

Moreover, we further explored the feasibility of other Pd_2X_3 monolayer phases, with X to be S and Te, respectively. Bulk PdS_2 has the same geometrical structure as that of bulk $PdSe_2$, thus the Pd_2S_3 monolayer might be experimentally synthesized following the same synthetic method as that of the Pd_2Se_3 monolayer. However, bulk $PdTe_2$ prefers a 1T configuration, indicating that the Pd_2Te_3 monolayer might be inaccessible. To confirm our assumption, we calculated the phonon dispersions of the two structures. No imaginary mode exists in the phonon spectra of Pd_2S_3 (see Figure 5a), indicating its dynamical stability of the monolayer. Whereas the phonon spectra of the Pd_2Te_3 monolayer shows imaginary frequency near the Γ point (see Figure S5), demonstrating its structural instability. We then calculated the electronic band structure of the stable Pd_2S_3 monolayer (Figure 5a), verifying the feature of a semiconductor with an indirect bandgap of 1.48 eV.

Since 2D TMCs can be vertically stacked layer-by-layer forming the van der Waals heterostructures which can efficiently modulate properties of materials for applications in nanoscale electronic and photovoltaic devices, here we propose a van der Waals heterostructure composed of the Pd_2Se_3 and Pd_2S_3 monolayers (see Figure 5b) and study its interesting properties. A key indicator for heterostructures is the band alignment that defines the type of heterostructures. Thus, we calculated the band alignment of the Pd_2Se_3 and Pd_2S_3 monolayers, as shown in Figure 5c. One can see that the Pd_2S_3/Pd_2Se_3 heterostructure has a type-II (ladder) band alignment, which allows more efficient electron-hole separation for lighting harvesting. Such type-II heterostructure can be used as active materials in excitonic solar cells (XSCs) [46–49]. For the Pd_2S_3/Pd_2Se_3 heterostructure, the Pd_2Se_3 monolayer is the donor and the Pd_2S_3 monolayer serves as the acceptor. With the approximation that the HSE06 bandgap equals optical bandgap and using the model developed by Scharber et al. [44], we obtained the upper limit of the PCE, reaching as high as 20% (Figure 5d). For comparison, we also calculated the band alignments for the $PdS_2/PdSe_2$ and $MoS_2/MoSe_2$ heterostructures, and find that they both are type-II heterostructures with predicted PCEs to be 14% and 12% respectively. The high PCE of the Pd_2S_3/Pd_2Se_3 heterostructure renders it a promising candidate in flexible optoelectronic and photovoltaic devices.

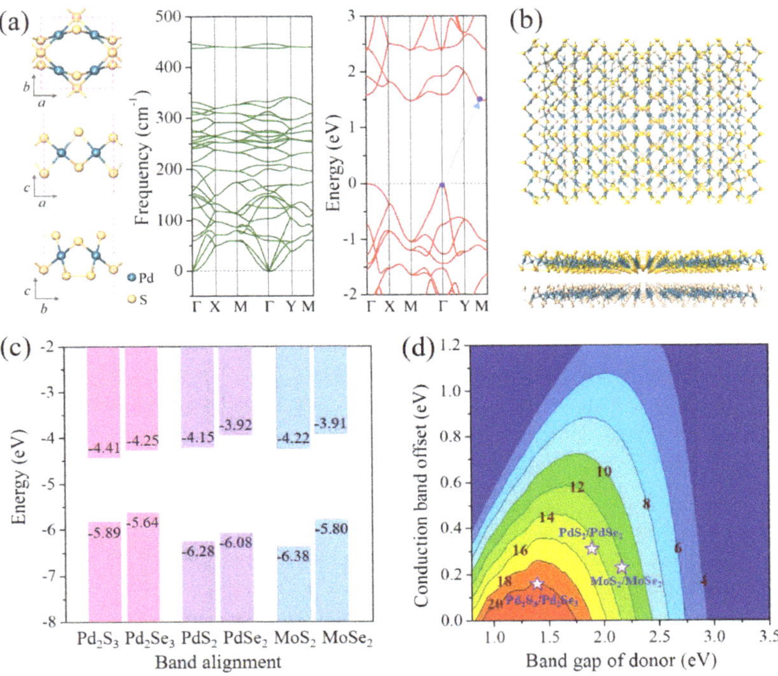

Figure 5. (**a**) Optimized atomic structure, phonon spectra and electronic band structure (at the HSE06 level) of the Pd_2S_3 monolayer; (**b**) Top and side views of the heterostructure composed of the Pd_2S_3 and Pd_2Se_3 monolayers; (**c**) Band alignments of the Pd_2S_3, Pd_2Se_3, PdS_2, $PdSe_2$, MoS_2, and $MoSe_2$ monolayers calculated using the HSE06 functional. The numbers are the CBM and VBM energies with respect to the vacuum level, which is set to zero when calculating the band alignment diagrams; (**d**) Computed PCE contour as a function of the donor bandgap and conduction band offset. Violet open stars mark the PCEs of Pd_2S_3/Pd_2Se_3, $PdS_2/PdSe_2$, and $MoS_2/MoSe_2$ heterostructure solar cells.

4. Conclusions

In summary, on the basis of DFT calculations, we systematically studied the properties and potential applications of the recently synthesized Pd_2Se_3 monolayer by focusing on its geometric structure, electronic band structure, and optical adsorption. Comparing with the previously reported $PdSe_2$ monolayer, we found that the Pd_2Se_3 monolayer has the following merits: (1) A suitable quasi-direct bandgap (1.39 eV) for light absorption, (2) a higher electron mobility (140.4 $cm^2V^{-1}s^{-1}$) and (3) a stronger optical absorption (~10^5 cm^{-1}) in the visible solar spectrum, showing promise of Pd_2Se_3 as an absorber material for future ultrathin photovoltaic devices. In addition, the Pd_2Se_3 monolayer combining with the stable Pd_2S_3 monolayer can form a type-II heterostructure, and the heterostructure solar cell system can achieve a 20% PCE. These findings would encourage experimentalists to devote more effort in developing Pd_2Se_3-based devices with high performance.

Supplementary Materials: The following are available online at http://www.mdpi.com/2079-4991/8/10/832/s1, Table S1: Structural parameters of the Pd_2Se_3, Pd_2S_3, Pd_2Te_3, PdS_2, $PdSe_2$, MoS_2, and $MoSe_2$ monolayers; Figure S1: (a) Raman spectra of $PdSe_2$ and Pd_2Se_3 monolayers. (b) and (c) are the corresponding Raman-active vibrational modes of the two structures; Figure S2: Electronic band structure of the Pd_2Se_3 monolayer calculated at the HSE06 level with and without considering the SOC interaction; Figure S3: Geometric structure of the Pd_2Se_3 monolayer under 0% and 9% biaxial tensile strain; Figure S4: Phonon dispersion of the Pd_2Se_3 monolayer under 9% biaxial tensile strain; Figure S5: Phonon dispersion of the Pd_2Te_3 monolayer.

Author Contributions: Investigation, X.L.; Methodology, X.L., S.Z. and Y.G.; Supervision, Q.W.; Writing—original draft, X.L.; Writing—review & editing, X.L., S.Z., Y.G., F.Q.W. and Q.W.

Funding: This work is partially supported by grants from the National Key Research and Development Program of China (2016YFE0127300, and 2017YFA0205003), the National Natural Science Foundation of China (NSFC-51471004, and NSFC-21773004). S. Z. is supported by the National Postdoctoral Program for Innovative Talents of China (BX201600091) and the Funding from China Postdoctoral Science Foundation (2017M610858).

Acknowledgments: This work is also supported by the High Performance Computing Platform of Peking University, China.

Conflicts of Interest: The authors declare no conflict of interest.

References

1. Lee, Y.H.; Zhang, X.Q.; Zhang, W.; Chang, M.T.; Lin, C.T.; Chang, K.D.; Yu, Y.C.; Wang Jacob, T.W.; Chang, C.S.; Li, L.J.; et al. Synthesis of Large-Area MoS_2 Atomic Layers with Chemical Vapor Deposition. *Adv. Mater.* **2012**, *24*, 2320–2325. [CrossRef] [PubMed]
2. Lu, X.; Utama, M.I.B.; Lin, J.; Gong, X.; Zhang, J.; Zhao, Y.; Pantelides, S.T.; Wang, J.; Dong, Z.; Liu, Z.; et al. Large-Area Synthesis of Monolayer and Few-Layer $MoSe_2$ Films on SiO_2 Substrates. *Nano Lett.* **2014**, *14*, 2419–2425. [CrossRef] [PubMed]
3. Xu, Z.-Q.; Zhang, Y.; Lin, S.; Zheng, C.; Zhong, Y.L.; Xia, X.; Li, Z.; Sophia, P.J.; Fuhrer, M.S.; Cheng, Y.-B.; et al. Synthesis and Transfer of Large-Area Monolayer WS_2 Crystals: Moving toward the Recyclable Use of Sapphire Substrates. *ACS Nano* **2015**, *9*, 6178–6187. [CrossRef] [PubMed]
4. Huang, J.-K.; Pu, J.; Hsu, C.-L.; Chiu, M.-H.; Juang, Z.-Y.; Chang, Y.-H.; Chang, W.-H.; Iwasa, Y.; Takenobu, T.; Li, L.-J. Large-Area Synthesis of Highly Crystalline WSe_2 Monolayers and Device Applications. *ACS Nano* **2014**, *8*, 923–930. [CrossRef] [PubMed]
5. Lin, Y.-C.; Komsa, H.-P.; Yeh, C.-H.; Björkman, T.; Liang, Z.-Y.; Ho, C.-H.; Huang, Y.-S.; Chiu, P.-W.; Krasheninnikov, A.V.; Suenaga, K. Single-Layer ReS_2: Two-Dimensional Semiconductor with Tunable in-Plane Anisotropy. *ACS Nano* **2015**, *9*, 11249–11257. [CrossRef] [PubMed]
6. Zhao, Y.; Qiao, J.; Yu, P.; Hu, Z.; Lin, Z.; Lau, S.P.; Liu, Z.; Ji, W.; Chai, Y. Extraordinarily Strong Interlayer Interaction in 2D Layered PtS_2. *Adv. Mater.* **2016**, *28*, 2399–2407. [CrossRef] [PubMed]
7. Sun, J.; Shi, H.; Siegrist, T.; Singh, D.J. Electronic, Transport, and Optical Properties of Bulk and Mono-Layer $PdSe_2$. *Appl. Phys. Lett.* **2015**, *107*, 153902. [CrossRef]
8. Oyedele, A.D.; Yang, S.; Liang, L.; Puretzky, A.A.; Wang, K.; Zhang, J.; Yu, P.; Pudasaini, P.R.; Ghosh, A.W.; Liu, Z.; et al. $PdSe_2$: Pentagonal Two-Dimensional Layers with High Air Stability for Electronics. *J. Am. Chem. Soc.* **2017**, *139*, 14090–14097. [CrossRef] [PubMed]
9. Feng, J.; Sun, X.; Wu, C.; Peng, L.; Lin, C.; Hu, S.; Yang, J.; Xie, Y. Metallic Few-Layered VS_2 Ultrathin Nanosheets: High Two-Dimensional Conductivity for in-Plane Supercapacitors. *J. Am. Chem. Soc.* **2011**, *133*, 17832–17838. [CrossRef] [PubMed]
10. Zhao, S.; Hotta, T.; Koretsune, T.; Watanabe, K.; Taniguchi, T.; Sugawara, K.; Takahashi, T.; Shinohara, H.; Kitaura, R. Two-Dimensional Metallic NbS_2: Growth, Optical Identification and Transport Properties. *2D Mater.* **2016**, *3*, 025027. [CrossRef]
11. Voiry, D.; Yang, J.; Chhowalla, M. Recent Strategies for Improving the Catalytic Activity of 2D TMD Nanosheets toward the Hydrogen Evolution Reaction. *Adv. Mater.* **2016**, *28*, 6197–6206. [CrossRef] [PubMed]
12. Radisavljevic, B.; Radenovic, A.; Brivio, J.; Giacometti, V.; Kis, A. Single-Layer MoS_2 Transistors. *Nat. Nanotechnol.* **2011**, *6*, 147. [CrossRef] [PubMed]
13. Akinwande, D.; Petrone, N.; Hone, J. Two-Dimensional Flexible Nanoelectronics. *Nat. Commun.* **2014**, *5*, 5678. [CrossRef] [PubMed]
14. Wang, H.; Yu, L.; Lee, Y.-H.; Shi, Y.; Hsu, A.; Chin, M.L.; Li, L.-J.; Dubey, M.; Kong, J.; Palacios, T. Integrated Circuits Based on Bilayer MoS_2 Transistors. *Nano Lett.* **2012**, *12*, 4674–4680. [CrossRef] [PubMed]
15. Yin, Z.; Li, H.; Li, H.; Jiang, L.; Shi, Y.; Sun, Y.; Lu, G.; Zhang, Q.; Chen, X.; Zhang, H. Single-Layer MoS_2 Phototransistors. *ACS Nano* **2012**, *6*, 74–80. [CrossRef] [PubMed]

16. Lopez-Sanchez, O.; Lembke, D.; Kayci, M.; Radenovic, A.; Kis, A. Ultrasensitive Photodetectors Based on Monolayer MoS_2. *Nat. Nanotechnol.* **2013**, *8*, 497. [CrossRef] [PubMed]
17. Mak, K.F.; He, K.; Shan, J.; Heinz, T.F. Control of Valley Polarization in Monolayer MoS_2 by Optical Helicity. *Nat. Nanotechnol.* **2012**, *7*, 494. [CrossRef] [PubMed]
18. Cao, T.; Wang, G.; Han, W.; Ye, H.; Zhu, C.; Shi, J.; Niu, Q.; Tan, P.; Wang, E.; Liu, B.; et al. Valley-Selective Circular Dichroism of Monolayer Molybdenum Disulphide. *Nat. Commun.* **2012**, *3*, 887. [CrossRef] [PubMed]
19. Soulard, C.; Rocquefelte, X.; Petit, P.E.; Evain, M.; Jobic, S.; Itié, J.P.; Munsch, P.; Koo, H.J.; Whangbo, M.H. Experimental and Theoretical Investigation on the Relative Stability of the PdS_2- and Pyrite-Type Structures of $PdSe_2$. *Inorg. Chem.* **2004**, *43*, 1943–1949. [CrossRef] [PubMed]
20. Chow, W.L.; Yu, P.; Liu, F.; Hong, J.; Wang, X.; Zeng, Q.; Hsu, C.H.; Zhu, C.; Zhou, J.; Wang, X.; et al. High Mobility 2D Palladium Diselenide Field-Effect Transistors with Tunable Ambipolar Characteristics. *Adv. Mater.* **2017**, *29*, 1602969. [CrossRef] [PubMed]
21. Lebègue, S.; Björkman, T.; Klintenberg, M.; Nieminen, R.M.; Eriksson, O. Two-Dimensional Materials from Data Filtering and Ab Initio Calculations. *Phys. Rev. X* **2013**, *3*, 031002.
22. Wang, Y.; Li, Y.; Chen, Z. Not Your Familiar Two Dimensional Transition Metal Disulfide: Structural and Electronic Properties of the PdS_2 Monolayer. *J. Mater. Chem. C* **2015**, *3*, 9603–9608. [CrossRef]
23. Lin, J.; Zuluaga, S.; Yu, P.; Liu, Z.; Pantelides, S.T.; Suenaga, K. Novel Pd_2Se_3 Two-Dimensional Phase Driven by Interlayer Fusion in Layered $PdSe_2$. *Phys. Rev. Lett.* **2017**, *119*, 016101. [CrossRef] [PubMed]
24. Zhu, X.; Li, F.; Wang, Y.; Qiao, M.; Li, Y. Pd_2Se_3 Monolayer: A Novel Two-Dimensional Material with Excellent Electronic, Transport, and Optical Properties. *J. Mater. Chem. C* **2018**, *6*, 4494–4500. [CrossRef]
25. Naghavi, S.S.; He, J.; Xia, Y.; Wolverton, C. Pd_2Se_3 Monolayer: A Promising Two-Dimensional Thermoelectric Material with Ultralow Lattice Thermal Conductivity and High Power Factor. *Chem. Mater.* **2018**, *30*, 5639–5647. [CrossRef]
26. Sebastian, Z.; Junhao, L.; Kazu, S.; Sokrates, T.P. Two-Dimensional $PdSe_2$-Pd_2Se_3 Junctions Can Serve as Nanowires. *2D Mater.* **2018**, *5*, 035025.
27. Perdew, J.P.; Burke, K.; Ernzerhof, M. Generalized Gradient Approximation Made Simple. *Phys. Rev. Lett.* **1996**, *77*, 3865–3868. [CrossRef] [PubMed]
28. Heyd, J.; Scuseria, G.E.; Ernzerhof, M. Hybrid Functionals Based on a Screened Coulomb Potential. *J. Chem. Phys.* **2003**, *118*, 8207–8215. [CrossRef]
29. Heyd, J.; Scuseria, G.E.; Ernzerhof, M. Erratum: "Hybrid Functionals Based on a Screened Coulomb Potential" [J. Chem. Phys. 118, 8207 (2003)]. *J. Chem. Phys.* **2006**, *124*, 219906. [CrossRef]
30. Blöchl, P.E. Projector Augmented-Wave Method. *Phys. Rev. B* **1994**, *50*, 17953–17979. [CrossRef]
31. Kresse, G.; Furthmüller, J. Efficient Iterative Schemes for Ab Initio Total-Energy Calculations Using a Plane-Wave Basis Set. *Phys. Rev. B* **1996**, *54*, 11169–11186. [CrossRef]
32. Monkhorst, H.J.; Pack, J.D. Special Points for Brillouin-Zone Integrations. *Phys. Rev. B* **1976**, *13*, 5188–5192. [CrossRef]
33. Parlinski, K.; Li, Z.Q.; Kawazoe, Y. First-Principles Determination of the Soft Mode in Cubic ZrO_2. *Phys. Rev. Lett.* **1997**, *78*, 4063–4066. [CrossRef]
34. Togo, A.; Tanaka, I. First Principles Phonon Calculations in Materials Science. *Scripta Mater.* **2015**, *108*, 1–5. [CrossRef]
35. Bardeen, J.; Shockley, W. Deformation Potentials and Mobilities in Non-Polar Crystals. *Phys. Rev.* **1950**, *80*, 72–80. [CrossRef]
36. Qiao, J.; Kong, X.; Hu, Z.-X.; Yang, F.; Ji, W. High-Mobility Transport Anisotropy and Linear Dichroism in Few-Layer Black Phosphorus. *Nat. Commun.* **2014**, *5*, 4475. [CrossRef] [PubMed]
37. Cai, Y.; Zhang, G.; Zhang, Y.-W. Polarity-Reversed Robust Carrier Mobility in Monolayer MoS_2 Nanoribbons. *J. Am. Chem. Soc.* **2014**, *136*, 6269–6275. [CrossRef] [PubMed]
38. Dai, J.; Zeng, X.C. Titanium Trisulfide Monolayer: Theoretical Prediction of a New Direct-Gap Semiconductor with High and Anisotropic Carrier Mobility. *Angew. Chem. Int. Ed.* **2015**, *54*, 7572–7576. [CrossRef] [PubMed]
39. Jing, Y.; Ma, Y.; Li, Y.; Heine, T. GeP_3: A Small Indirect Band Gap 2d Crystal with High Carrier Mobility and Strong Interlayer Quantum Confinement. *Nano Lett.* **2017**, *17*, 1833–1838. [CrossRef] [PubMed]
40. Saha, S.; Sinha, T.P.; Mookerjee, A. Electronic Structure, Chemical Bonding, and Optical Properties of Paraelectric $BaTiO_3$. *Phys. Rev. B* **2000**, *62*, 8828–8834. [CrossRef]

41. Miao, N.; Xu, B.; Bristowe, N.C.; Zhou, J.; Sun, Z. Tunable Magnetism and Extraordinary Sunlight Absorbance in Indium Triphosphide Monolayer. *J. Am. Chem. Soc.* **2017**, *139*, 11125–11131. [CrossRef] [PubMed]
42. Wang, B.; Niu, X.; Ouyang, Y.; Zhou, Q.; Wang, J. Ultrathin Semiconducting Bi_2Te_2S and Bi_2Te_2Se with High Electron Mobilities. *J. Phys. Chem. Lett.* **2018**, *9*, 487–490. [CrossRef] [PubMed]
43. Paier, J.; Marsman, M.; Kresse, G. Dielectric Properties and Excitons for Extended Systems from Hybrid Functionals. *Phys. Rev. B* **2008**, *78*, 121201. [CrossRef]
44. Scharber, M.C.; Mühlbacher, D.; Koppe, M.; Denk, P.; Waldauf, C.; Heeger, A.J.; Brabec, C.J. Design Rules for Donors in Bulk-Heterojunction Solar Cells-Towards 10 % Energy-Conversion Efficiency. *Adv. Mater.* **2006**, *18*, 789–794. [CrossRef]
45. Bernardi, M.; Palummo, M.; Grossman, J.C. Semiconducting Monolayer Materials as a Tunable Platform for Excitonic Solar Cells. *ACS Nano* **2012**, *6*, 10082–10089. [CrossRef] [PubMed]
46. Dai, J.; Zeng, X.C. Bilayer Phosphorene: Effect of Stacking Order on Bandgap and Its Potential Applications in Thin-Film Solar Cells. *J. Phys. Chem. Lett.* **2014**, *5*, 1289–1293. [CrossRef] [PubMed]
47. Ganesan, V.D.S.O.; Linghu, J.; Zhang, C.; Feng, Y.P.; Shen, L. Heterostructures of Phosphorene and Transition Metal Dichalcogenides for Excitonic Solar Cells: A First-Principles Study. *Appl. Phys. Lett.* **2016**, *108*, 122105. [CrossRef]
48. Xie, M.; Zhang, S.; Cai, B.; Huang, Y.; Zou, Y.; Guo, B.; Gu, Y.; Zeng, H. A Promising Two-Dimensional Solar Cell Donor: Black Arsenic–Phosphorus Monolayer with 1.54 eV Direct Bandgap and Mobility Exceeding 14,000 $cm^2v^{-1}s^{-1}$. *Nano Energy* **2016**, *28*, 433–439. [CrossRef]
49. Lee, J.; Huang, J.; Sumpter, B.G.; Yoon, M. Strain-Engineered Optoelectronic Properties of 2D Transition Metal Dichalcogenide Lateral Heterostructures. *2D Mater.* **2017**, *4*, 021016. [CrossRef]
50. Shockley, W.; Queisser, H.J. Detailed Balance Limit of Efficiency of P-N Junction Solar Cells. *J. Appl. Phys.* **1961**, *32*, 510–519. [CrossRef]
51. Hanna, M.C.; Nozik, A.J. Solar Conversion Efficiency of Photovoltaic and Photoelectrolysis Cells with Carrier Multiplication Absorbers. *J. Appl. Phys.* **2006**, *100*, 074510. [CrossRef]
52. Wadia, C.; Alivisatos, A.P.; Kammen, D.M. Materials Availability Expands the Opportunity for Large-Scale Photovoltaics Deployment. *Environ. Sci. Technol.* **2009**, *43*, 2072–2077. [CrossRef] [PubMed]
53. Motta, C.; El-Mellouhi, F.; Kais, S.; Tabet, N.; Alharbi, F.; Sanvito, S. Revealing the Role of Organic Cations in Hybrid Halide Perovskite $CH_3NH_3PbI_3$. *Nat. Commun.* **2015**, *6*, 7026. [CrossRef] [PubMed]
54. Zheng, F.; Tan, L.Z.; Liu, S.; Rappe, A.M. Rashba Spin–Orbit Coupling Enhanced Carrier Lifetime in $CH_3NH_3PbI_3$. *Nano Lett.* **2015**, *15*, 7794–7800. [CrossRef] [PubMed]
55. Zhang, C.; Sun, Q. A Honeycomb BeN_2 Sheet with a Desirable Direct Bandgap and High Carrier Mobility. *J. Phys. Chem. Lett.* **2016**, *7*, 2664–2670. [CrossRef] [PubMed]

Article

Strain-Tunable Visible-Light-Responsive Photocatalytic Properties of Two-Dimensional CdS/g-C_3N_4: A Hybrid Density Functional Study

Guangzhao Wang [1,*], Feng Zhou [1], Binfang Yuan [2,*], Shuyuan Xiao [3], Anlong Kuang [4,*], Mingmin Zhong [4], Suihu Dang [1], Xiaojiang Long [1] and Wanli Zhang [1]

1 School of Electronic Information Engineering, Key Laboratory of Extraordinary Bond Engineering and Advanced Materials Technology of Chongqing, Yangtze Normal University, Chongqing 408100, China; zhoufeng9966@126.com (F.Z.); dangsuihu@126.com (S.D.); longxiaojiang@yznu.edu.cn (X.L.); zhangwl@yznu.cn (W.Z.)

2 School of Chemistry and Chemical Engineering, Yangtze Normal University, Chongqing 408100, China

3 Institute for Advanced Study, Nanchang University, Nanchang 330031, China; syxiao@hust.edu.cn

4 School of Physical Science and Technology, Southwest University, Chongqing 400715, China; zhongmm@swu.edu.cn

* Correspondence: wangyan6930@yznu.edu.cn or wangyan6930@126.com (G.W.); 6781022@163.com (B.Y.); alkuang@swu.edu.cn (A.K.)

Received: 9 January 2019; Accepted: 5 February 2019; Published: 12 February 2019

Abstract: By means of a hybrid density functional, we comprehensively investigate the energetic, electronic, optical properties, and band edge alignments of two-dimensional (2D) CdS/g-C_3N_4 heterostructures by considering the effect of biaxial strain and pH value, so as to improve the photocatalytic activity. The results reveal that a CdS monolayer weakly contacts with g-C_3N_4, forming a type II van der Waals (vdW) heterostructure. The narrow bandgap makes CdS/g-C_3N_4 suitable for absorbing visible light and the induced built-in electric field between the interface promotes the effective separation of photogenerated carriers. Through applying the biaxial strain, the interface adhesion energy, bandgap, and band edge positions, in contrast with water, redox levels of CdS/g-C_3N_4 can be obviously adjusted. Especially, the pH of electrolyte also significantly influences the photocatalytic performance of CdS/g-C_3N_4. When pH is smaller than 6.5, the band edge alignments of CdS/g-C_3N_4 are thermodynamically beneficial for oxygen and hydrogen generation. Our findings offer a theoretical basis to develop g-C_3N_4-based water-splitting photocatalysts.

Keywords: CdS/g-C_3N_4; strain-tunable; photocatalysis; water splitting; hybrid density functional

1. Introduction

Gaining hydrogen through photocatalytic water splitting by use of solar energy provides a new way to solve the problems of energy shortage and environmental pollution. A large number of semiconductors, such as TiO_2 [1], ZnO [2], $KNbO_3$ [3], and $NaNbO_3$ [4] have drawn much attention as promising photocatalysts, but they can merely utilize ultraviolet light, which only makes up only 4% of solar energy. Some photocatalysts, such as bulk CdS [5], have suitable bandgaps for visible light absorption, but lacks stability due to the self-oxidation of photogenerated species. Thus, it is challenging to find efficient water-splitting photocatalysts, and some appropriate strategies should be taken to modulate the electronic and photocatalytic properties of pristine photocatalysts. Generally, introduction of dopants [6,7], loading noble metal [8], dye sensitizing [9] and cocatalysis through constructing heterojunctions [10–12] are effective at improving the photocatalytic activity. The desired photocatalyst must have the conduction band minimum (CBM) and valence band maximum (VBM) individually above the water reduction (H^+/H_2) potential and below the water oxidation (O_2/H_2O)

potential. Besides, the theoretical minimum bandgap of 1.23 eV is required for water splitting [13] considering the overpotential accompanied by water redox processes.

Since graphene was prepared, 2D materials including hexagonal boron nitride [14], graphite-like zinc oxide [15], transition-metal dichalcogenides [16], and MXenes [17] have been extensively investigated and utilized in the area of optoelectronics and photocatalysts. Particularly, the graphite-like carbon nitride (g-C_3N_4) is a prospective photocatalyst used for hydrogen generation by photocatalytic decomposition of water [18]. g-C_3N_4 has a suitable bandgap of 2.7 eV for visible light absorption. However, g-C_3N_4 exhibits poor photocatalytic efficiency because of the fast recombination of photogenerated electron–hole pairs [19–21]. This factor obviously restrains the photocatalytic efficiency of g-C_3N_4. It is of great significance to adopt measures to regulate the electronic structures of g-C_3N_4 in a bid to enhance the photocatalytic performance. Especially, a large number of 2D heterostructures, such as ZnO/WS_2 [22], AlN/WS_2 [23], GaN/WS_2 [24], g-C_6N_6/g-C_3N_4 [25], g-C_3N_4/MoS_2 [26] and g-C_3N_4/C_2N [27] exhibit significantly improved photocatalytic activity as compared to pristine 2D materials. In these heterostructures, the formed built-in electric field caused by the charge accumulation/depletion around the interfaces promotes the effective separation and migration of photogenerated carriers, which is beneficial to enhance the photocatalytic performance. A recent theoretical study [28] reports the stability, electronic structures, and offset of 2D CdS/g-C_3N_4 heterostructure, and the result suggests that the heterostructure has suitable bandgap and band alignments for visible light photocatalytic water splitting. Moreover, the induced electric field between CdS layer and g-C_3N_4 also accelerates the separation of photogenerated carriers and improves the photocatalytic activity. However, whether the biaxial strain will improve the photocatalytic activity of CdS/g-C_3N_4 is still unclear. Besides, it is also unclear whether the photocatalytic activity of CdS/g-C_3N_4 is affected by the pH of electrolyte. These two problems have to be solved in a bid to obviously enhance the photocatalytic performance of CdS/g-C_3N_4.

The purpose of this work is to investigate the energetic, electronic, optical property and band edge alignments of CdS/g-C_3N_4 as well as the effect induced by the biaxial strain and the pH of electrolyte, in order to regulate the photocatalytic performance. This work is organized as follows. Section 2 depicts the computational details, while Section 3 displays the results and discussion about the energetic, optical, optical, band edge alignments as well as the photocatalytic property of CdS/g-C_3N_4 heterostructure with the consideration of biaxial strain and pH, and ultimately Section 4 lists some concluding remarks.

2. Computational Details

The CdS/g-C_3N_4 heterostructure, which consists of 3 Cd, 3 S, 6 C, and 8 N atoms, is constructed through vertically stacking a $\sqrt{3} \times \sqrt{3}$ supercell of hexagonal CdS single-layer on a 1×1 g-C_3N_4 cell. We carry out density functional theory (DFT) calculations by means of the general gradient approximation (GGA) [29] of Perdew–Burke–Ernzerhof (PBE) [30] and hybrid density functional of HSE06 [31], as implemented in the Vienna ab initio simulation package (VASP) [32]. We adopt the projected-augmented-wave (PAW) method [33] to describe the electron-ion interaction and DFT-D3 correction [34] to well treat long-range vdW interaction. To avoid the interactions introduced by the periodic structures, a vacuum of 18 Å is used. We first optimize the geometries by use of PBE, and then accurately calculate the electronic and optical properties by utilization of HSE06. The plane-wave cutoff energy is set as 500 eV, and a Monkhorst-pack [35] k-point mesh of $13 \times 13 \times 1$ for CdS cell, $9 \times 9 \times 1$ for g-C_3N_4 cell and CdS/g-C_3N_4 heterostructures are used. All the structures are fully relaxed until the energy and force on each atom are individually reduced to 10^{-5} eV and 0.02 eV/Å. The valence electron configurations of of Cd ($4d^{10}5s^2$), S ($3s^23p^4$), C($2s^22p^2$), and N ($2s^22p^3$) are considered to construct the PAW potentials.

Finally, the optical absorption spectra of g-C_3N_4 and CdS/g-C_3N_4 composite is calculated by use of HSE06. The absorption coefficient is obtained from the the real and imaginary parts of the frequency dependent complex dielectric function $\varepsilon(\omega)$=$\varepsilon_1(\omega)$+$i\varepsilon_2(\omega)$ according to the following relationship [36]

$$I(\omega) = \sqrt{2}\omega\sqrt{\sqrt{\varepsilon_1^2(\omega)+\varepsilon_2^2(\omega)}-\varepsilon_1(\omega)} \tag{1}$$

The imaginary part of the dielectric function ε_2 is calculated as [37]

$$\varepsilon_2(\hbar\omega) = \frac{2e^2\pi}{\Omega\varepsilon_0}\sum_{k,v,c}|\langle\psi_k^c|\mathbf{u}\cdot\mathbf{r}|\psi_k^v\rangle|^2\delta(E_k^c - E_k^v - \hbar\omega) \tag{2}$$

where Ω, v, c, ω, $\mathbf{u}$, ψ_k^v and ψ_k^c denotes the unit-cell volume, valence bands, conduction bands, photon frequencies, the vector defining the polarization of the incident electric field, the occupied and unoccupied wave functions at point k in reciprocal space, respectively, while the real part of the dielectric function ε_1 can be obtained from imaginary part ε_2 by the Kramer-Kronig relationship [38].

$$\varepsilon_1(\omega) = 1 + \frac{2}{\pi}p\int_0^\infty \frac{\varepsilon_2(\omega')\omega'}{\omega'^2-\omega^2}d\omega' \tag{3}$$

where p denotes the principal value of the integral.

3. Results and Discussion

The geometry structures, density of states (DOS) and projected density of states (PDOS) of CdS monolayer and g-C_3N_4 are depicted in Figure 1. The calculated lattice constants for CdS and g-C_3N_4 single-layers are respectively a = b = 4.245 and a = b = 7.134 Å, and the obtained bandgaps for CdS and g-C_3N_4 single-layers are, respectively, 2.74 and 2.77 eV, which are well consistent with previous experiment and theoretical reports [28]. The VBM of CdS single-layer mainly consists of S 3p, Cd 4d and Cd 4p orbitals, whereas the CBM is primarily contributed by Cd 5s character. For g-C_3N_4, the VBM is mainly composed of N 2p orbitals with some amount of C 2p and N 2s orbitals, while the CBM is comprised of C 2p and N 2p characters.

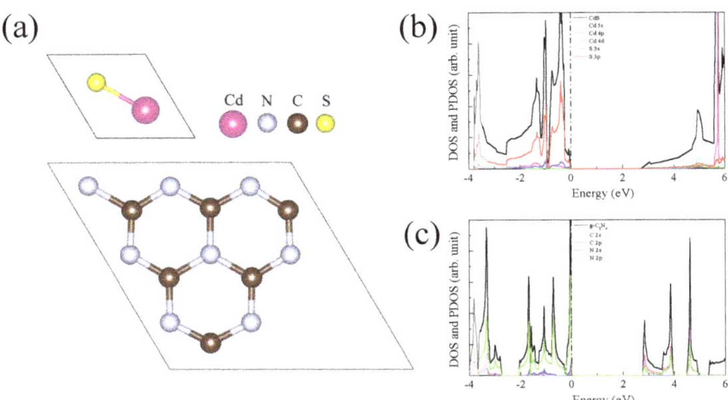

Figure 1. (**a**) Crystal structures of CdS single-layer and g-C_3N_4. DOS and PDOS of (**b**) CdS single-layer and (**c**) g-C_3N_4.

The lattice mismatch is defined as: $[(L_{g-C_3N_4} - L_{s-CdS})/L_{s-CdS}] \times 100\%$, where $L_{g-C_3N_4}$ and L_{s-CdS} are the lattice constants of g-C_3N_4 cell and $\sqrt{3}\times\sqrt{3}$ CdS supercell, respectively. When a $\sqrt{3}\times\sqrt{3}$ CdS supercell contacts with a 1×1 g-C_3N_4 cell, the lattice mismatch is only -2.97%, which is good for the construction of CdS/g-C_3N_4 heterostructure. We consider a $\sqrt{3} \times \sqrt{3}$ CdS supercell with with tree special rotation angles of 0°, 120°, and 240° sitting on a 1×11 g-C_3N_4 cell with fixed angles to construct three possible configurations of CdS/g-C_3N_4, as depicted in Figure 2. These different heterostructures are call as CdS/g-C_3N_4 (i), (ii), and (iii), respectively. The optimized lattice constants for CdS/g-C_3N_4 (i), (ii) and (iii) are respectively 6.954, 6.955 and 6.920 Å, slightly smaller than the lattice of g-C_3N_4. This may be attributed to the atom rearrangements in the heterostructures. The obtained bandgaps for CdS/g-C_3N_4 (i), (ii) and (iii) are 2.745, 2.746 and 2.676 eV, respectively. Though the bandgaps of these heterostructures are almost the same as the bandgap of g-C_3N_4, the absorption of visible light is significantly improved. This will be detailed in the following discussion.

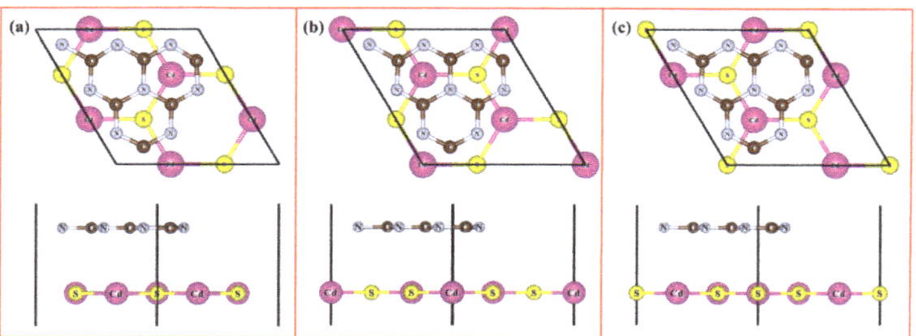

Figure 2. Top and side views of three possible stackings of CdS/g-C_3N_4 heterostructures.

To explore the thermodynamic stability, the interface binding energy (E_b) is calculated according to the following relationship:

$$E_b = E_{CdS/g\text{-}C_3N_4} - E_{CdS} - E_{g\text{-}C_3N_4} \tag{4}$$

where $E_{CdS/g\text{-}C_3N_4}$, E_{CdS}, and $E_{g\text{-}C_3N_4}$ denote the total energies of CdS/g-C_3N_4 heterostructure, CdS single-layer, and g-C_3N_4, respectively. The E_b values for CdS/g-C_3N_4 (i), (ii) and (iii) are respectively −1.62633, −1.62548 and −1.62630 eV, implying these heterostructures are exothermic and are energetically favorable. Besides, the differences of E_b among these structures are so small that these three configurations may be experimentally prepared at the same time. These three configurations have similar energy values. Furthermore, the band alignments depicted in Figure 3 also indicate that the band edge positions of these three heterostructures are close. Thus, our discussion is mainly focused on the CdS/g-C_3N_4 (i). The interface adhesion energy (E_a) is calculated according to the following equation:

$$E_a = E_b/S \tag{5}$$

where S is the area of CdS/g-C_3N_4 heterostructure vertical to the vacuum direction. The E_a for CdS/g-C_3N_4 (i) is -19.4 meV/Å^2, within the scope of typical vdW heterostructure of -20 meV/Å^2 [39].

As an ideal water-splitting photocatalyst, its band edges must be located in proper positions. The CBM and VBM must straddle the water redox potentials to satisfy the thermodynamic criterion for overall water splitting. Figure 3 displays the band edge alignments for CdS monolayer, g-C_3N_4, CdS/g-C_3N_4 (i), (ii) and (iii). The band edges of these systems are all straddle the water redox levels, which is propitious to spontaneously produce both hydrogen and oxygen.

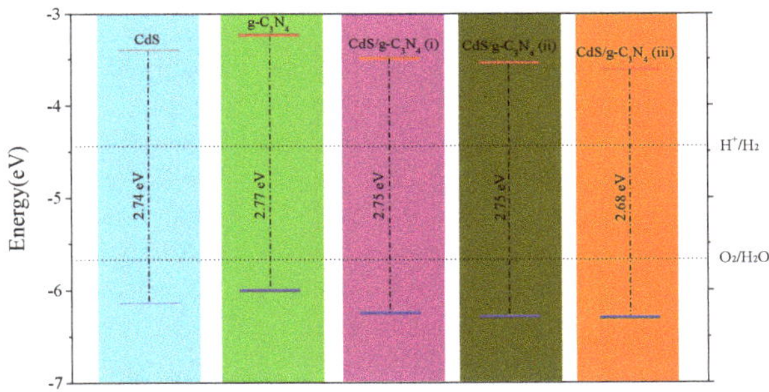

Figure 3. Band edge alignments for CdS single-layer, g-C_3N_4, CdS/g-C_3N_4 (i), (ii), and (iii) in contrast with water redox levels.

The appearance of strain can not be ignored due to the lattice mismatch between different 2D semiconductors. It is found that for 2D materials, the electronic and optical properties can be modulated through strain engineering [40–42]. We consider the influence caused by both tensile and compressive biaxial strain on the energetic, electronic, and photocatalytic properties of CdS/g-C_3N_4. The biaxial strain is defined as $\epsilon = [(a - a_0)/a_0] \times 100\%$, in which a and a_0 are the lattice parameters of strained and pristine CdS/g-C_3N_4 heterostructures, respectively. $\epsilon < 0$ means the structure is compressed, while $\epsilon > 0$ means the structure is stretched. Figure 4 gives the varied E_a and E_g values of CdS/g-C_3N_4 heterostructures of different biaxial strain with 2% apart. The E_a value gets smaller within the scope of $\epsilon = -8\%$ to $\epsilon = 0$ but gets larger in the range of $\epsilon = 0$ to $\epsilon = 8\%$. The unstrained CdS/g-C_3N_4 heterostructure has the least interface adhesion energy, which implies that unstrained configuration has advantage in energy in contrast with strained configuration. The calculated E_a value with the ϵ in the range from −8% to 8% are 82.2, 55.1, 7.1, −11.8, −19.4, −11.7, 6.8, 32.9 and 52.8 meV/Å^2, indicating the formation of the heterostructures with the ϵ = −2%, 0, 2% are exothermic. The E_g value decreases in the range of $\epsilon = -8\%$ to $\epsilon = -6\%$, increases in the range of $\epsilon = -6\%$ to $\epsilon = 0$, and decreases in the range of $\epsilon = 0$ to $\epsilon = 8\%$. This suggests that the visible light absorption can be modulated by tuning the bandgaps through biaxial strain engineering. The unstrained heterostructure has the largest bandgap. The obtained bandgaps for CdS/g-C_3N_4 heterostructures in the range of $\epsilon = -8\%$ to $\epsilon = 8\%$ are 2.43, 0.78, 1.72, 2.19, 2.75, 2.54, 2.34, 2.20 and 1.34 eV.

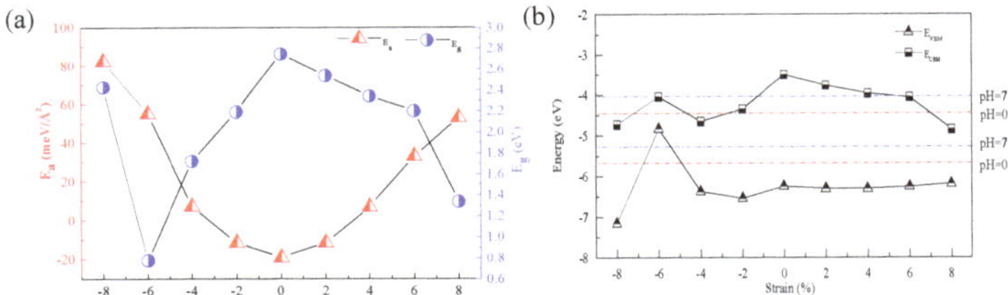

Figure 4. (**a**) Varied interface adhesion energies (E_a) and bandgaps (E_g) of CdS/g-C_3N_4 heterostructures with different biaxial strains. (**b**) Band edge alignments of CdS/g-C_3N_4 heterostructures with different biaxial strains. The red and blue horizontal lines are the water redox potentials as pH = 0 and pH = 7, respectively.

The photocatalytic performance is affected by the pH of electrolyte. Particularly, the standard hydrogen electrode potential varies with the pH varies. The standard reduction (H^+/H_2) in contrast with the vacuum level is calculated by: $E_{H^+/H_2} = -4.44$ eV + pH $\times$ 0.059 eV [43]. With the consideration of the difference of 1.23 eV [44] between water redox potentials during the water redox reactions, the oxygen potential (O_2/H_2O) is calculated by: $E_{O_2/H_2O} = E_{H^+/H_2} - 1.23$ eV =

−5.67 eV + pH × 0.059 eV. The method has been successfully applied to predict the photocatalytic properties of P and As doped C_2N monolayer [45], CdS/ZnSe heterostructure [46], and (Bule P)/BSe heterostructure [47] with considering the effect of pH on the standard redox potentials with respect to the vacuum level.

The band edge alignments of CdS/g-C_3N_4 heterostructures with diverse biaxial strains are displayed in Figure 4. Both CBM and VBM of CdS/g-C_3N_4 of ϵ = −2%, 0, 2%, 4%, 6% individually straddle the water redox levels in the pH range of 0–1.6, 0–14, 0–11.5, 0–8.1, 0–6.5. In the pH range of 0–14, the VBM and CBM of CdS/g-C_3N_4 with ϵ = −8%, −4%, 8% are individually lower than the water oxidation (O_2/H_2O) and reduction (H^+/H_2) potentials, which means that these heterostructures are only beneficial for oxygen generation. For the case of ϵ = −6%, the VBM and CBM are individually above the water oxidation (O_2/H_2O) and reduction (H^+/H_2) potentials when the pH is lower than 6.9. When the pH is lower than 6.5, the CdS/g-C_3N_4 with ϵ = 0, 2%, 4%, 6% are thermodynamically feasible for over all water redox reactions, while the composites of ϵ = −2%, −6%, −8%, −4%, 8% are propitious to spontaneously generate oxygen. Therefore, adjustment of the pH lower than 6.5 is conducive to improve the photocatalytic activity of CdS/g-C_3N_4.

Next, we plot the DOS, PDOS, and band structures of unstrained CdS/g-C_3N_4 to shed light on the physical mechanism of water splitting of CdS/g-C_3N_4. Figure 5 shows that the CBM and VBM are individually donated by g-C_3N_4 and CdS layer, suggesting that the CdS/g-C_3N_4 is a type II heterostructure. The partial charge density of CBM and VBM in Figure 6 also suggests the CBM of g-C_3N_4 is predominately contributed by g-C_3N_4 and the VBM is mainly donated by CdS layer. The VBM is primarily composed of S 3p, Cd 4d and Cd 4p states, while the CBM is predominately dominated by N 2p and C 2p states. Taking the electronic transition of angular momentum selection rules of $\triangle l = \pm 1$ into account, after absorbing photons, the electrons primarily migrate from Cd 4d orbitals below the Fermi level to N 2p and C 2p orbitals in conduction band.

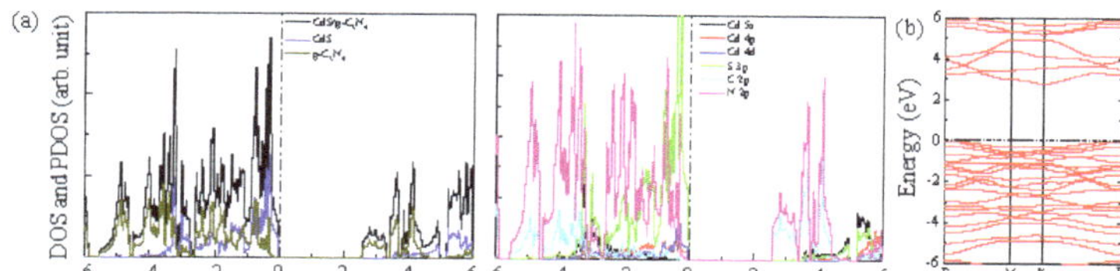

Figure 5. (**a**) DOS, PDOS and (**b**) band structures of CdS/g-C_3N_4 heterostructure.

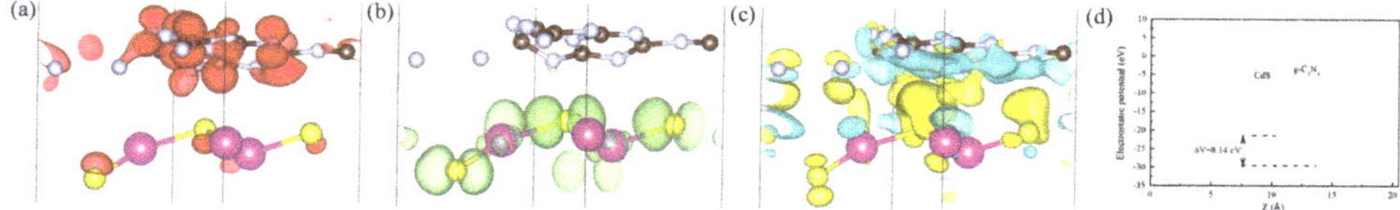

Figure 6. Partial charge densities of (**a**) CBM, (**b**) VBM, (**c**) the charge density difference, and (**d**) potential drop across the interface of CdS/g-C_3N_4 heterostructure.

The charge density difference of CdS/g-C_3N_4 heterostructure in Figure 6c, where cyan and yellow regions represent charge depletion and accumulation, respectively. It is obvious that electrons migrate from CdS layer to g-C_3N_4. Based on the Bader charge analysis, the transferred charge is 0.027 $|e|$, which is enough to introduce a large potential drop between the g-C_3N_4 and CdS layer. Figure 6d lists potential drop across the interface of CdS/g-C_3N_4 along the Z direction, i.e., the vacuum direction. The g-C_3N_4 has a deeper potential as compared to that of CdS layer, which drives electrons to migrate from CdS layer to g-C_3N_4. The potential drop (ΔV) across the interface is 8.14 eV, inducing a built-in electric field from the g-C_3N_4 to CdS layer. The formed built-in electric field can promote the shifts of photogenerated carriers, thus further inhibiting the recombination of photogenerated carriers. The g-C_3N_4 and CdS individually pose as electron acceptor and donor. Thus, the water oxidation reaction and reduction reaction occur on the CdS layer and g-C_3N_4, respectively. This is beneficial for improving the photocatalytic activity.

Another key indicator to the photocatalytic performance is the optical absorption. Figure 7 depicts the obtained absorption cures for g-C_3N_4 and CdS/g-C_3N_4, the original g-C_3N_4 only exhibits a obvious absorption above 3.0 eV and there is almost no visible light absorption ability for g-C_3N_4, which may be due to the fact that only a small amount of electron density migrates electron migrates from N 2s states of valence band to C 2p and N 2p states of conduction band (see Figure 1). The adsorption edge of CdS/g-C_3N_4 shifts to 2.7 eV, especially the g-C_3N_4 shows stronger light absorption than g-C_3N_4 in the range of 2.7–4.3 eV, i.e., the CdS/g-C_3N_4 owns a broad absorption in both ultraviolet and visible light regions. According to Figures 1 and 5, the reason of enhancement of light absorption should be that the electron migration from the Cd 4d states below the Fermi level to C 2p and N 2p states are significantly enhanced as compared to pristine g-C_3N_4.

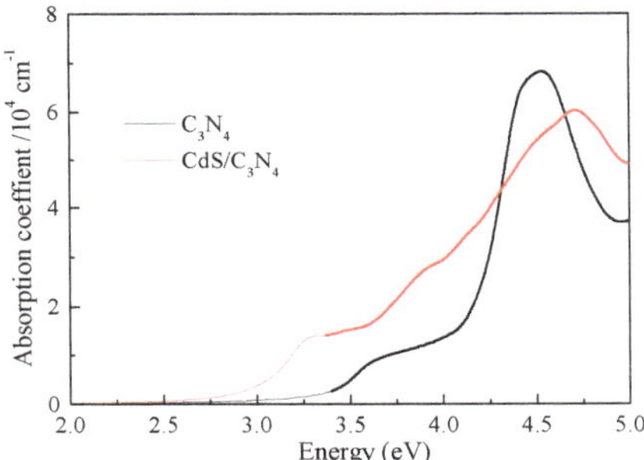

Figure 7. Absorption spectra of pristine g-C_3N_4 and CdS/g-C_3N_4 heterostructure.

4. Conclusions

In summary, the hybrid density functional HSE06 is employed to calculate the energetic, electronic and optical properties of CdS/g-C_3N_4, whilst taking into account different biaxial strains as well as the pH of electrolyte, in a bid to tune the photocatalytic activity of CdS/g-C_3N_4. When the interaction between single-layer CdS and g-C_3N_4, the vdW CdS/g-C_3N_4 heterostructure is easy to form, as the interface adhesion formation energy is negative. The predicted bandgaps and optical absorptions indicate the CdS/g-C_3N_4 heterostructure can absorb visible light. Furthermore, the formed built-in electric field around the interface region is helpful to accelerate electron–hole recombination. The bandgaps, interface adhesion energies, and band edge alignments in reference to water redox potentials are visibly affected by the biaxial strains. The photocatalytic performance of CdS/g-C_3N_4 can be modulated by tuning the biaxial strains and pH. When pH is lower than 6.5, the band edge positions of CdS/g-C_3N_4 are thermodynamically favorable for spontaneously producing of oxygen and hydrogen. In general, CdS/g-C_3N_4 is a perspective water-splitting photocatalyst.

Author Contributions: G.W., B.Y., and A.K. designed the project, guided the study, and prepared the manuscript; F.Z., S.X., and M.Z. carried the calculations; S.D., X.L. and W.Z. analyzed the calculated results and produced the illustrations.

Funding: This work was supported by the National Natural Science Foundation of China under Grant Nos. 11504301 and 11847100, and the Science and Technology Plan Project of Fuling District under Grant No. FLKJ,2018BBA3056.

Acknowledgments: The work was carried out at LvLiang Cloud Computing Center of China, and the calculations were performed on TianHe-2.

Conflicts of Interest: The authors declare no conflict of interest.

References

1. Yin, W.; Wei, S.; Aljassim, M.M.; Yan, Y. Double-hole-mediated coupling of dopants and its impact on band gap engineering in TiO_2. *Phys. Rev. Lett.* **2011**, *106*, 066801. [CrossRef] [PubMed]
2. Pan, J.; Wang, S.; Chen, Q.; Hu, J.; Wang, J. Band-structure engineering of ZnO by anion-cation co-doping for enhanced photo-electrochemical activity. *ChemPhysChem* **2014**, *15*, 1611–1618. [CrossRef] [PubMed]
3. Wang, G.; Chen, H.; Li, Y.; Kuang, A.; Yuan, H.; Wu, G. A hybrid density functional study on the visible light photocatalytic activity of (Mo,Cr)-N codoped $KNbO_3$. *Phys. Chem. Chem. Phys.* **2015**, *17*, 28743–28753. [PubMed]
4. Wang, G.; Chen, H.; Wu, G.; Kuang, A.; Yuan, H. Hybrid density functional study on mono- and codoped $NaNbO_3$ for visible-light photocatalysis. *ChemPhysChem* **2016**, *17*, 489–499. [CrossRef] [PubMed]
5. Liu, J. Origin of high photocatalytic efficiency in monolayer g-C_3N_4/CdS heterostructure: a hybrid DFT study. *J. Phys. Chem. C* **2015**, *119*, 28417–28423. [CrossRef]
6. Wang, G.; Huang, Y.; Kuang, A.; Yuan, H.; Li, Y.; Chen, H. Double-hole-mediated codoping on $KNbO_3$ for visible light photocatalysis. *Inorg. Chem.* **2016**, *55*, 9620–9631. [CrossRef] [PubMed]
7. Gai, Y.; Li, J.; Li, S.; Xia, J.; Wei, S. Design of narrow-gap TiO_2: A passivated codoping approach for enhanced Pphotoelectrochemical activity. *Phys. Rev. Lett.* **2009**, *102*, 036402. [CrossRef]
8. Meir, N.; Plante, I.J.; Flomin, K.; Chockler, E.; Moshofsky, B.; Diab, M.; Volokh, M.; Mokari, T. Studying the chemical, optical and catalytic properties of noble metal (Pt, Pd, Ag, Au)-Cu_2O core-shell nanostructures grown via a general approach. *J. Mater. Chem.* **2013**, *1*, 1763–1769. [CrossRef]
9. Leijtens, T.; Lim, J.; Teuscher, J.; Park, T.; Snaith, H.J. Charge density dependent mobility of organic hole-transporters and mesoporous TiO_2 determined by transient mobility spectroscopy: implications to dye-sensitized and organic solar cells. *Adv. Mater.* **2013**, *25*, 3227–3233. [CrossRef]
10. Yang, W.; Wen, Y.; Zeng, D.; Wang, Q.; Chen, R.; Wang, W.; Shan, B. Interfacial charge transfer and enhanced photocatalytic performance for the heterojunction WO_3/BiOCl: First-principles study. *J. Mater. Chem.* **2014**, *2*, 20770–20775. [CrossRef]
11. Torabi, A.; Staroverov, V.N. Band gap reduction in ZnO and ZnS by creating layered ZnO/ZnS heterostructures. *J. Phys. Chem. Lett.* **2015**, *6*, 2075–2080. [CrossRef] [PubMed]
12. Cui, X.; Yang, X.; Xian, X.; Tian, L.; Tang, H.; Liu, Q. Insights into highly improved solar-driven photocatalytic oxygen evolution over integrated Ag_3PO_4/MoS_2 heterostructures. *Front. Chem.* **2018**, *6*, 123. [CrossRef] [PubMed]
13. Liao, J.; Sa, B.; Zhou, J.; Ahuja, R.; Sun, Z. Design of high-efficiency visible-light photocatalysts for water splitting: MoS_2/AlN(GaN) heterostructures. *J. Phys. Chem. C* **2014**, *118*, 17594–17599. [CrossRef]
14. Golberg, D.; Bando, Y.; Huang, Y.; Terao, T.; Mitome, M.; Tang, C.; Zhi, C. Boron nitride nanotubes and nanosheets. *ACS Nano* **2010**, *4*, 2979–2993. [CrossRef] [PubMed]
15. Wang, G.; Yuan, H.; Chang, J.; Wang, B.; Kuang, A.; Chen, H. ZnO/MoX_2 (X = S, Se) composites used for visible light photocatalysis. *RSC Adv.* **2018**, *8*, 10828–10835. [CrossRef]
16. Chhowalla, M.; Shin, H.S.; Eda, G.; Li, L.; Loh, K.P.; Zhang, H. The chemistry of two-dimensional layered transition metal dichalcogenide nanosheets. *Nat. Chem.* **2013**, *5*, 263–275. [CrossRef] [PubMed]
17. He, J.; Ding, G.; Zhong, C.; Li, S.; Li, D.; Zhang, G. Cr_2TiC_2-based double MXenes: Novel 2D bipolar antiferromagnetic semiconductor with gate-controllable spin orientation toward antiferromagnetic spintronics. *Nanoscale* **2019**, *11*, 356–364. [CrossRef]
18. Wang, X.; Maeda, K.; Thomas, A.; Takanabe, K.; Xin, G.; Carlsson, J.M.; Domen, K.; Antonietti, M. A metal-free polymeric photocatalyst for hydrogen production from water under visible light. *Nat. Chem.* **2009**, *8*, 76–80. [CrossRef]
19. Zhang, M.; Yao, W.; Lv, Y.; Bai, X.; Zhu, Y. Enhancement of mineralization ability of C_3N_4 via a lower valence position by a tetracyanoquinodimethane organic semiconductor. *J. Mater. Chem. A* **2014**, *2*, 11432–11438. [CrossRef]
20. Liang, Y.; Long, C.; Li, J.; Huang, B.; Dai, Y. InSe monolayer: Promising cocatalyst of g-C_3N_4 for water splitting under Visible Light. *ACS Appl. Energy Mater.* **2018**, *1*, 5394–5401. [CrossRef]
21. Liu, J.; Liu, Y.; Liu, N.; Han, Y.; Zhang, X.; Huang, H.; Lifshitz, Y.; Lee, S.; Zhong, J.; Kang, Z. Metal-free efficient photocatalyst for stable visible water splitting via a two-electron pathway. *Science* **2015**, *347*, 970–974. [CrossRef]

22. Wang, G.; Li, D.; Sun, Q.; Dang, S.; Zhong, M.; Xiao, S.; Liu, G. Hybrid density functional study on the photocatalytic properties of two-dimensional g-ZnO based heterostructures. *Nanomaterials* **2018**, *8*, 374. [CrossRef]
23. Wang, G.; Dang, S.; Zhang, P.; Xiao, S.; Wang, C.; Zhong, M. Hybrid density functional study on the photocatalytic properties of AlN/$MoSe_2$, AlN/WS_2, and AlN/WSe_2 heterostructures. *J. Phys. D Appl. Phys.* **2018**, *51*, 025109. [CrossRef]
24. Wang, G.; Dang, S.; Zhao, W.; Li, Y.; Xiao, S.; Zhong, M. Tunable photocatalytic properties of GaN-based two-dimensional heterostructures. *Phys. Status Solidi B* **2018**. [CrossRef]
25. Liang, D.; Jing, T.; Ma, Y.; Hao, J.; Sun, G.; Deng, M. Photocatalytic properties of g-C_6N_6/g-C_3N_4 heterostructure: a theoretical study. *J. Phys. Chem. C* **2016**, *120*, 24023–24029. [CrossRef]
26. Li, J.; Liu, E.; Ma, Y.; Hu, X.; Wan, J.; Sun, L.; Fan, J. Synthesis of MoS_2/g-C_3N_4 nanosheets as 2D heterojunction photocatalysts with enhanced visible light activity. *Appl. Surf. Sci.* **2016**, *364*, 694–702. [CrossRef]
27. Wang, H.; Li, X.; Yang, J. The g-C_3N_4/C_2N nanocomposite: A g-C_3N_4-based water-splitting photocatalyst with enhanced energy efficiency. *ChemPhysChem* **2016**, *17*, 2100–2104. [CrossRef]
28. Li, J.; Wei, W.; Mu, C.; Huang, B.; Dai, Y. Electronic properties of van der Waals g-C_3N_4/CdS heterojunction from first-principles. *Phys. E* **2018**, *103*, 459–463. [CrossRef]
29. Perdew, J.P.; Burke, K.; Ernzerhof, M. Generalized gradient approximation made simple. *Phys. Rev. Lett.* **1996**, *77*, 3865. [CrossRef]
30. Ernzerhof, M.; Scuseria, G.E. Assessment of the Perdew-Burke-Ernzerhof exchange-correlation functional. *J. Chem. Phys.* **1999**, *110*, 5029–5036. [CrossRef]
31. Heyd, J.; Scuseria, G.E.; Ernzerhof, M. Hybrid functionals based on a screened Coulomb potential. *J. Chem. Phys.* **2003**, *118*, 8207–8215. [CrossRef]
32. Kresse, G.; Furthmüller, J. Efficient iterative schemes for ab initio total-energy calculations using a plane-wave basis set. *Phys. Rev. B* **1996**, *54*, 11169. [CrossRef]
33. Blöchl, P.E. Projector augmented-wave method. *Phys. Rev. B* **1994**, *50*, 17953. [CrossRef]
34. Grimme, S.; Antony, J.; Ehrlich, S.; Krieg, H. A consistent and accurate ab initio parametrization of density functional dispersion Ccorrection (DFT-D) for the 94 elements H-Pu. *J. Chem. Phys.* **2010**, *132*, 154104. [CrossRef]
35. Monkhorst, H.J.; Pack, J.D. Special points for Brillouin-zone integrations. *Phys. Rev. B* **1976**, *13*, 5188–5192. [CrossRef]
36. Saha, S.; Sinha, T.P.; Mookerjee, A. Electronic structure, chemical bonding, and optical properties of paraelectric $BaTiO_3$. *Phys. Rev. B* **2000**, *62*, 8828–8834. [CrossRef]
37. Tian, F.; Liu, C. DFT description on electronic structure and optical absorption properties of anionic S-doped anatase TiO_2. *J. Phys. Chem. B* **2006**, *110*, 17866–17871. [CrossRef]
38. Fu, Q.; Li, J.L.; He, T.; Yang, G.W. Band-engineered $CaTiO_3$ nanowires for visible light photocatalysis. *J. Appl. Phys.* **2013**, *113*, 37. [CrossRef]
39. Bjorkman, T.; Gulans, A.; Krasheninnikov, A.V.; Nieminen, R.M. Van der Waals bonding in layered compounds from advanced density-functional first-principles calculations. *Phys. Rev. Lett.* **2012**, *108*, 235502. [CrossRef]
40. Scalise, E.; Houssa, M.; Pourtois, G.; Ev, V.A.; Stesmans, A. Strain-induced semiconductor to metal transition in the two-dimensional honeycomb structure of MoS_2. *Nano Res.* **2012**, *5*, 43–48. [CrossRef]
41. Yu, W.; Zhu, Z.; Zhang, S.; Cai, X.; Wang, X.; Niu, C.; Zhang, W. Tunable electronic properties of GeSe/phosphorene heterostructure from first-principles study. *Appl. Phys. Lett.* **2016**, *109*, 103104. [CrossRef]
42. Li, S.; Wang, C.; Qiu, H. Single- and few-layer ZrS_2 as efficient photocatalysts for hydrogen production under visible light. *Int. J. Hydrogen Energy* **2015**, *40*, 15503–15509. [CrossRef]
43. Chakrapani, V.; Angus, J.C.; Anderson, A.B.; Wolter, S.D.; Stoner, B.R.; Sumanasekera, G. Charge transfer equilibria between diamond and an aqueous oxygen electrochemical redox couple. *Science* **2007**, *318*, 1424–1430. [CrossRef]
44. Artrith, N.; Sailuam, W.; Limpijumnong, S.; Kolpak, A.M. Reduced overpotentials for electrocatalytic water splitting over Fe- and Ni-modified $BaTiO_3$. *Phys. Chem. Chem. Phys.* **2016**, *18*, 29561–29570. [PubMed]
45. Kishore, M.R.A.; Ravindran, P. Tailoring the electronic band gap and band edge positions in the C_2N Monolayer by P and As substitution for photocatalytic water splitting. *J. Phys. Chem. C* **2017**, *121*, 22216–22224. [CrossRef]

46. Yang, H.; Li, J.; Yu, L.; Huang, B.; Ma, Y.; Ying, D. Theoretical study on electronic properties of in-plane CdS/ZnSe heterostructure: Type-II band alignment for water splitting. *J. Mater. Chem. A* **2018**, *6*, 4161–4166. [CrossRef]
47. Wang, B.; Li, X.; Zhao, R.; Cai, X.; Yu, W.; Li, W.; Liu, Z.; Zhang, L.; Ke, S. Electronic structures and enhanced photocatalytic properties of blue phosphorene/BSe van der Waals heterostructures. *J. Mater. Chem. A* **2018**, *6*, 8923–8929. [CrossRef]

Article

Hybrid Density Functional Study on the Photocatalytic Properties of Two-dimensional g-ZnO Based Heterostructures

Guangzhao Wang [1], Dengfeng Li [2,*], Qilong Sun [3], Suihu Dang [1], Mingmin Zhong [4,*], Shuyuan Xiao [5] and Guoshuai Liu [6]

1 School of Electronic Information Engineering, Yangtze Normal University, Chongqing 408100, China; wangyan6930@126.com (G.W.); dangsuihu@126.com (S.D.)
2 Department of Science, Chongqing University of Posts and Telecommunications, Chongqing 400065, China
3 Chongqing Institute of Green and Intelligent Technology, Chinese Academy of Sciences, Chongqing 400714, China; sunqilong@cigit.ac.cn
4 School of Physical Science and Technology, Southwest University, Chongqing 400715, China
5 Wuhan National Laboratory for Optoelectronics, Huazhong University of Science and Technology, Wuhan 430074, China; syxiao@hust.edu.cn
6 State Key Laboratory of Urban Water Resource and Environment, School of Environment, Harbin Institute of Technology, Harbin 150090, China; swift_ft@163.com
* Correspondence: lidf@cqupt.edu.cn (D.L.); zhongmm@swu.edu.cn (M.Z.); Tel.: +86-23-624-71346 (D.L.); +86-23-683-67040 (M.Z.)

Received: 11 May 2018; Accepted: 25 May 2018; Published: 28 May 2018

Abstract: In this work, graphene-like ZnO (g-ZnO)-based two-dimensional (2D) heterostructures (ZnO/WS_2 and ZnO/WSe_2) were designed as water-splitting photocatalysts based on the hybrid density functional. The dependence of photocatalytic properties on the rotation angles and biaxial strains were investigated. The bandgaps of ZnO/WS_2 and ZnO/WSe_2 are not obviously affected by rotation angles but by strains. The ZnO/WS_2 heterostructures with appropriate rotation angles and strains are promising visible water-splitting photocatalysts due to their appropriate bandgap for visible absorption, proper band edge alignment, and effective separation of carriers, while the water oxygen process of the ZnO/WSe_2 heterostructures is limited by their band edge positions. The findings pave the way to efficient g-ZnO-based 2D visible water-splitting materials.

Keywords: ZnO/WS_2; ZnO/WSe_2; photocatalysis; hybrid density functional

1. Introduction

An increasing amount of effort has been dedicated to 2D materials for their distinctive electronic [1], optical [2,3], mechanical properties [4], and their potential applications in superconductivity [5], supercapacitors [6], lithium-ion batteries [7], solar cells [8], and photocatalysis [9]. Recently, graphene-like ZnO (g-ZnO) has been experimentally synthesized [10,11] and proven to be energetically stable by density functional theory (DFT) [12]. Though there have been many investigations [13–16] focused on the magnetism of g-ZnO, few studies exist regarding the water-splitting [17,18] of g-ZnO. As bulk ZnO-based materials are promising water-splitting photocatalysts, we may wonder about the photocatalytic activity of g-ZnO- and g-ZnO-based materials. However, the bandgap for g-ZnO is 3.25 eV [13], which results in inefficient visible light absorption and reduces the utilization of solar energy. Therefore, the electronic structure of g-ZnO should be adjusted so as to reduce the bandgap and absorb more visible light. A desired water-splitting photocatalyst should have a conduction band minimum (CBM) and a valence band maximum (VBM) above the water reduction level and the water oxidation level, respectively [19,20]. Considering the additional overpotential accompanied with overall water redox processes, the theoretical bandgap for desired water-splitting photocatalyst should be larger than 1.23 eV [19,20]. Construction of a heterojunction is a useful method to improve the photocatalytic performance of photocatalysts [21–26]. The monolayer WS_2 (WSe_2) has been studied as a photocatalyst; the appropriate bandgap of 1.98 (1.63) [27] eV ensures its strong ability for visible light absorption. As monolayer WS_2 (WSe_2) has a similar crystal structure and almost the same lattice constants compared with g-ZnO, we consider building heterostructures between g-ZnO and the WS_2 (WSe_2) monolayer, i.e., ZnO/WS_2 (ZnO/WSe_2) heterostructures.

In this article, using the hybrid density functional, the structural, electronic, and optical properties and band edge alignment of ZnO/WS_2 and ZnO/WSe_2 heterostructures are described and discussed to explore whether they have an efficient visible light response and photocatalytic activities. The following questions are posed: (i) Will these two heterostructures be promising water-splitting photocatalysts with an appropriate bandgap and band edge positions? (ii) Will charge separation exist between the constituent monolayers? (iii) Will these two heterostructures have an efficient absorption of visible light? (iv) Will the electronic and optical properties be changed with the application of acceptable strains?

2. Computational Details

The heterostructure models of ZnO/WS_2 and ZnO/WSe_2 are built using a 2×2 supercell of g-ZnO as a substrate to support 2×2 supercells of WS_2 and WSe_2 monolayers, i.e., the lattice parameters of the heterostructures are the fixed value of optimized 2 × 2 g-ZnO of $a = b = 6.58$ Å. The calculated lattice mismatch between ZnO and WS_2 (WSe_2) monolayer is −3.4% (+0.3%), which is helpful for experimental preparations of ZnO/WS_2 and ZnO/WSe_2. In addition, a vacuum space of 18 Å is adopted to avoid the interactions between neighboring nonocomposites. The Vienna *ab initio* simulation package (VASP) [28] was used to perform the DFT calculations, and the Perdew-Burke-Ernzernof (PBE) [29] under generalized gradient approximation (GGA) [30] within the projected augmented wave (PAW) method [31] are utilized. The DFT-D3 [32] vdW correction by Grime is adopted to treat the weak van der Waals (vdW) interactions. An energy cutoff of 500 eV, energy convergence thresholds of 10^{-5} eV, force convergence criteria of 0.01 eV/Å, and k-points of 13 × 13 × 1 for 1 × 1 g-ZnO, WS_2 (WSe_2) monolayers and 7 × 7 × 1 for 2 × 2 ZnO/WS_2 (ZnO/WSe_2) are sufficient for calculating geometric and electronic structures. To determine electronic and optical properties more accurately, the hybrid density functional of Heyd-Scuseria-Ernzerhof [33,34] (HSE06) with a mixing coefficient of 0.25 is used. In summary, the PBE is used for structural optimizations and energy calculations, while the HSE06 is adopted for the calculation of electronic structures and optical properties. Furthermore, the valence states of O ($2s^22p^4$), S ($3s^23p^4$), Se ($4s^24p^4$), Zn ($3d^{10}4s^2$), and W ($5p^66s^25d^4$) are used to construct PAW potentials. The absorption curves are calculated from the imaginary part of the dielectric constant according to the Kramers-Kroning dispersion relation [35].

3. Results and Discussion

The obtained bandgaps for the g-ZnO and WS_2 (WSe_2) monolayers are, respectively, 3.30 and 2.35 (2.10) eV, consistent with previous reports [27,36]. The obtained lattice parameters for the g-ZnO and WS_2 monolayers are 3.290 and 3.180 (3.300) Å, respectively. The lattice mismatch between the g-ZnO and WS_2 WSe_2 of −3.4% (0.3%) is small, which is favorable for the construction of a ZnO/WS_2 (ZnO/WSe_2) heterostructure. To build the ZnO/WS_2 and ZnO/WSe_2 heterostructure models, six different ZnO single-layers rotating on the fixed WS_2 and WSe_2 monolayers from 0 to 300° with 60° apart are considered. Top views of different stacked ZnO/WS_2 and ZnO/WSe_2 heterostructures are depicted in Figure 1. The Zn–O bond lengths in all these ZnO/WS_2 and ZnO/WSe_2 heterostructures are the same value of 1.900 Å, which is easy to understand because the lattice parameters of these heterostructures are the fixed values of the 2 × 2 g-ZnO and WS_2 layer (WSe_2) hardly affects the ZnO layer in the composites because of the weak vdW interactions. The lengths of the W–S bond in ZnO/WS_2 with rotation angles in the range of 0–300° are, respectively, 2.441

(2.444), 2.440 (2.445), 2.442 (2.444), 2.444 (2.444), 2.444 (2.444), and 2.442 (2.444) Å, and the lengths of the W–Se bond in ZnO/WSe_2 are, respectively, 2.540 (2.544), 2.539 (2.544), 2.541 (2.544), 2.544 (2.544), 2.543 (2.544), and 2.542 (2.544). The length of the W–S (W–Se) bond in ZnO/WS_2 (ZnO/WSe_2) are slightly larger (smaller) than the original length of the W–S (W–Se) bond in the WS_2 (WSe_2) monolayer, which is due to the fact that a small lattice mismatch causes small atom rearrangements. When the rotation angles are in the range of 0–300°, the layer distances between the two layers in ZnO/WSe_2 (ZnO/WSe_2) are 2.976, 2.932, 2.964, 3.316, 3.328, and 2.974 (3.084, 3.071, 3.036, 3.375, 3.374, and 3.067) Å, respectively.

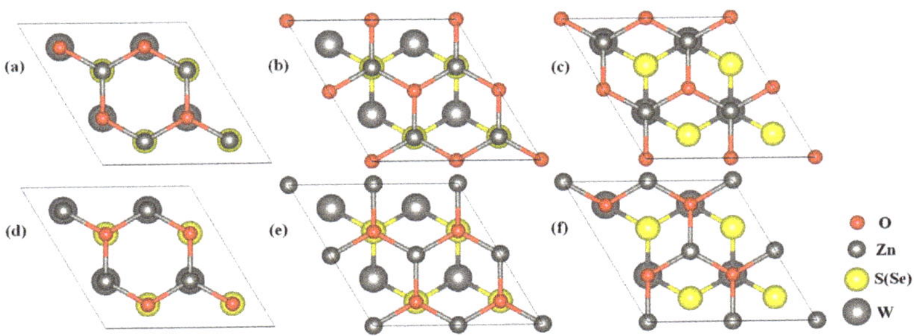

Figure 1. Top views of the ZnO/WS_2 (ZnO/WSe_2) with the g-ZnO in different rotation angles of (**a**) 0° (the reference); (**b**) 60°; (**c**) 120°; (**d**) 180°; (**e**) 240°; and (**f**) 300°.

The relative stability of ZnO/WS_2 and ZnO/WSe_2 could be compared through a calculation of interface adhesion energy. The interface adhesion energies (E_a) for ZnO/WS_2 (ZnO/WSe_2) are defined as

$$E_a = [E_{ZnO/WS_2(WSe_2)} - E_{ZnO} - E_{WS_2(WSe_2)}]/S \quad (1)$$

where $E_{ZnO/WS_2(WSe_2)}$, E_{ZnO}, and $E_{WS_2(WSe_2)}$ are the total energies for the relaxed ZnO/WS_2 (ZnO/WSe_2), g-ZnO, and WS_2 (WSe_2) monolayers. S is the top area of the heterostructure. Figure 2a gives the E_a values of ZnO/WS_2 and ZnO/WSe_2 of different rotation angles, and all these six configurations for both ZnO/WS_2 and ZnO/WSe_2 heterostructures possess negative interface adhesion energies, implying that the formation of these interfaces are exothermic and that these heterostructures could be easily prepared. It is interesting that the varied tendency of the E_a values for ZnO/WS_2 and ZnO/WSe_2 with different rotation angles are almost the same, which is attributed to the similar geometric structures and elemental compositions of these heterostructures. Either ZnO/WS_2 or ZnO/WSe_2 has a minimum E_a value at a rotation angle of 120° in the corresponding heterostructures, indicating that these two heterostructure configurations are the most stable structures in the considered configurations. When the rotation angle is 120°, the E_a values for ZnO/WS_2 and ZnO/WSe_2 are respectively −16.28 and −29.92 meV/Å^2, within the scope of a vdW E_a value of around 20 meV/Å^2 [37]. This indicates that ZnO/WS_2 and ZnO/WSe_2 are vdW heterostructures. Figure 2 shows the varied bandgaps of ZnO/WS_2 and ZnO/WSe_2 with different rotation angles. The calculated bandgaps for ZnO/WS_2 (ZnO/WSe_2) of the rotation angles in the range of 0–300° are 1.33, 1.35, 1.48, 1.487, 1.491, and 1.50 (2.14, 2.125, 2.134, 2.15, 2.14, and 2.16) eV, respectively. The bandgaps for ZnO/WS_2 and ZnO/WSe_2 heterostructures are obviously smaller than the bandgap of the ZnO monolayer and favorable for visible light absorption. The bandgaps of ZnO/WS_2 and ZnO/WSe_2 heterostructures are almost unchanged when the rotation angles vary, meaning that the rotation component has a negligible impact on the bandgaps of these heterostructures, i.e., the different stacked models will not qualitatively affect our conclusion. Therefore, we could neglect the tiny effect on the electronic structures of heterostructures caused by the rotational component. The following calculations and discussions about the effect of strains on the electronic structures are focused on the ZnO/WS_2 and ZnO/WSe_2 with the smallest E_a value, i.e., ZnO/WS_2 and ZnO/WSe_2 with the rotation angle of 120°.

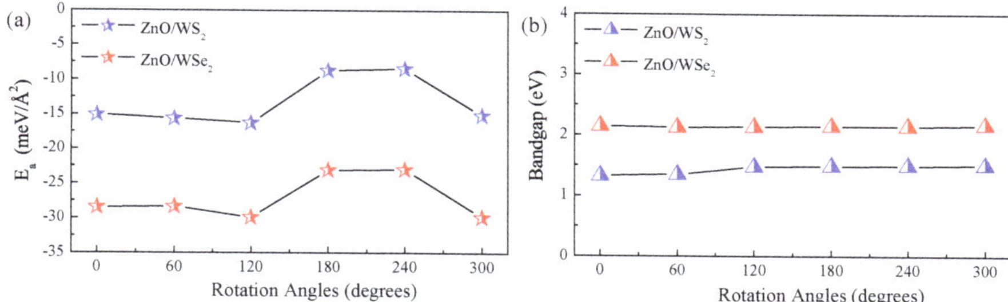

Figure 2. (**a**) Interface adhesion energies and (**b**) bandgaps of ZnO/WS_2 and ZnO/WSe_2 with different rotation angles.

The suitable bandgap may not always ensure the enhancement of photocatalytic activity. One should also pay attention to band edge alignment in reference to the water redox level. A desired water-splitting photocatalyst must have a VBM lower than the water oxidation level and a CBM higher than the water reduction level. Figure 3 plots the band edge alignment of ZnO/WS_2 and ZnO/WSe_2 of different rotation angles. The band edge positions of ZnO/WS_2 with the rotation angles of 120, 180, 240, and 300° straddle the water redox levels, suggesting that these heterostructures have the ability to act as photocatalysts for the overall water splitting process. For ZnO/WS_2 with rotation angles of 0 and 60°, the CBM positions are lower than the water reduction level, which make these two heterostructures unfavorable for the spontaneous production of hydrogen. For ZnO/WSe_2 with different rotation angles, VBM positions are above the water oxidation level, which causes poor oxygen evolution efficiency.

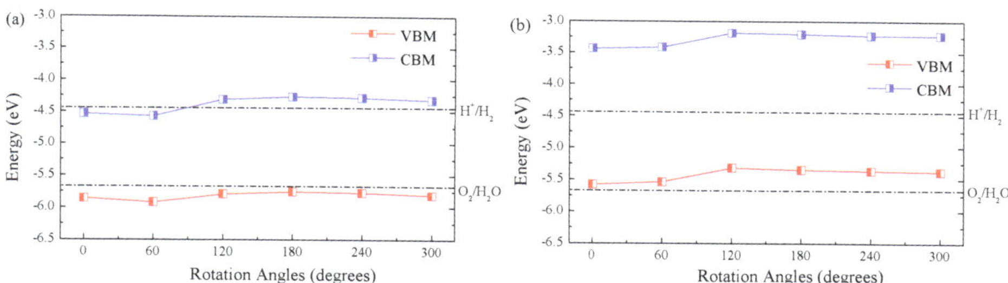

Figure 3. Band alignment of (**a**) ZnO/WS_2 and (**b**) ZnO/WSe_2 of different rotation angles with respect to the water redox levels.

When two materials with different lattice constants form a heterostructure, the strain will obviously affect the geometry and electronic properties. In addition, many studies report that the electronic and optical properties of 2D materials [38–40] could be effectively tuned through the application of strain. The biaxial strain, which is calculated as $\epsilon = [(a - a_0)/a_0] \times 100\%$ (a and a_0 are, respectively, the lattice parameters with and without biaxial strain), is considered to alter the photocatalytic activities of ZnO/WS_2 and ZnO/WSe_2. Figure 4a indicates that E_a values become smaller in the range of $\epsilon = -6\%$–-2% but become larger in the range of $\epsilon = -2\%$–$+6\%$, which means that the ZnO/WS_2 with a strain of -2% is a more stable configuration as compared to these others. The E_a values of ZnO/WSe_2 become smaller in the range of $\epsilon = -6\%$–0 but become larger in the range of ϵ = 0–+6%, implying that the ZnO/WSe_2 without strain is energetically more favorable in contrast with g-ZnO. The E_a values for ZnO/WS_2 of $\epsilon = -2\%$ and ZnO/WSe_2 of $\epsilon = 0$ are, respectively, −22.97 and −29.92 meV/Å^2. Hence, these two heterostructures belong to vdW heterostructures. The bandgaps of ZnO/WS_2 and ZnO/WSe_2 of different strains are depicted in Figure 4b. The bandgaps for ZnO/WS_2 of ϵ = −6%–+6% are, respectively, 1.61, 2.05, 1.94, 1.48, 1.04, 0.68, and 0.39 eV, and the bandgaps for ZnO/WSe_2 of ϵ = −6%–+6% are, respectively, 1.63, 1.95, 2.06, 2.13, 1.84, 1.60, and 1.22 eV. The bandgaps of ZnO/WS_2 become larger in the range of ϵ = −6%–−4% but become smaller in the range of ϵ = −4%–+6%, and the ZnO/WS_2 of $\epsilon = -4\%$ has the largest bandgap. The bandgaps of ZnO/WSe_2 become larger in the range of ϵ = −6%–0 but become smaller in the range of ϵ = 0–+6%, i.e., the ZnO/WSe_2 without strain has the largest bandgap.

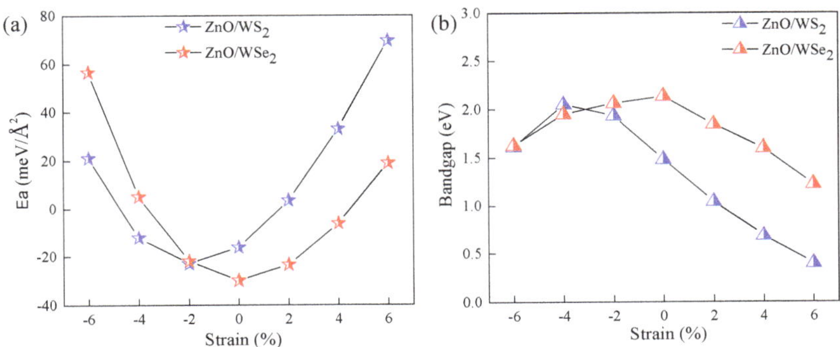

Figure 4. (**a**) Interface adhesion energies and (**b**) bandgaps of ZnO/WS_2 and ZnO/WSe_2 with different strains.

The band edge alignment of ZnO/WS_2 and ZnO/WSe_2 heterostructures with different strains is given in Figure 5. The band edge positions of ZnO/WS_2 heterostructures with $\epsilon = -2\%$ straddle the water redox levels, implying that these heterostructures are suitable for both hydrogen and oxygen evolution. For ZnO/WS_2 heterostructures with $\epsilon = -6\%$ and −4%, the CBM levels are suitable for hydrogen evolution, but VBM levels are unfavorable for oxygen evolution. While for the case of ZnO/WSe_2 heterostructure with $\epsilon = +2\%$, though the VBM level is favorable for spontaneous oxygen production, the CBM level is unfavorable for spontaneous hydrogen production. The band edge positions of ZnO/WS_2 heterostructures with $\epsilon = +4\%$ and +6% lie between the water reduction potential and water oxygen potential, which makes these heterostructures unfavorable for over all water splitting process. For the case of ZnO/WSe_2 heterostructures with different strains, though the CBM levels are suitable for generating hydrogen, the VBM levels are unfavorable for generating oxygen.

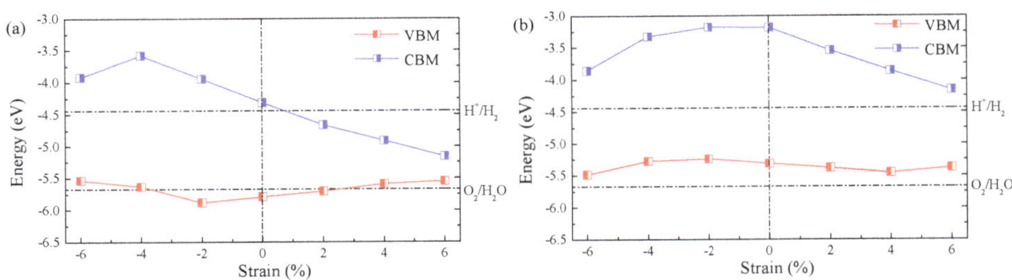

Figure 5. Band alignment of (**a**) ZnO/WS_2 and (**b**) ZnO/WSe_2 with different strains.

The DOS, PDOS, and band structures of ZnO/WS_2 and ZnO/WSe_2 are shown in Figure 6. The CBM and VBM, respectively, are located at K and Γ, suggesting that ZnO/WS_2 has an indirect bandgap. The CBM is primarily caused by W 5d orbitals and a small amount of S 3p orbitals, while VBM predominantly consists of W 5d orbitals. The electrons below the Fermi levels are mainly excited from W 5d (O 2p, S 3p) to S 3p (W 5d) orbitals, when the electronic transition of angular momentum selection rules of $\Delta l = \pm 1$ is considered. Figure 7a indicates the electrons in the ZnO layer will migrate to the WS_2 layer, which will be helpful for effective separation of photogenerated carriers. Both the CBM and VBM of ZnO/WSe_2 are prominently caused by W 5d and Se 4p orbitals and a small amount of O 2p orbitals. After absorbing the photo energy, the electrons in the W 5d (Se 4p) orbitals below the Fermi level will jump to W 5d (Se 4p) orbitals of the conduction band, and only a small amount of electrons jump from W 5d (O 2p) to O 2p (W 5d) orbitals. Figure 7b implies the electrons in the ZnO layer will transfer to the WSe_2 layer, which is usually favorable for the effective separation of photogenerated carriers.

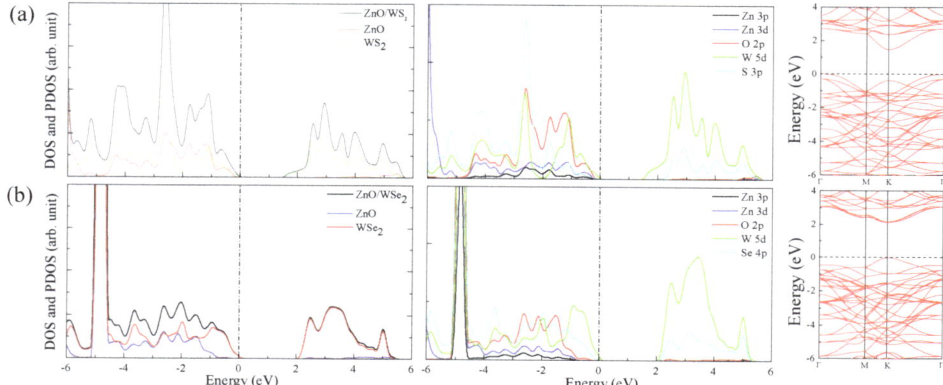

Figure 6. DOS, PDOS, and band structures of (**a**) ZnO/WS_2 and (**b**) ZnO/WSe_2.

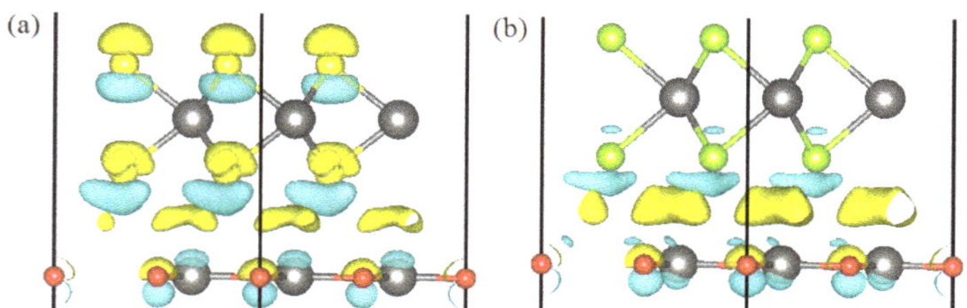

Figure 7. Side views of the charge differences of (**a**) ZnO/WS_2 and (**b**) ZnO/WSe_2.

The calculated optical absorption curves for the ZnO, WS_2, WSe_2 monolayers and the ZnO/WS_2 and ZnO/WSe_2 heterostructures are depicted in Figure 8. The absorption curve of g-ZnO is limited to the ultraviolet region, whereas WS_2 and WSe_2 monolayers could absorb visible light and show obvious visible light absorption. Moreover, it is noted that ZnO/WS_2 could absorb more visible light as compared to the g-ZnO and WS_2 monolayers. The visible light absorption of ZnO/WSe_2 is not improved in contrast with the WSe_2 monolayer but is obviously improved in contrast with g-ZnO.

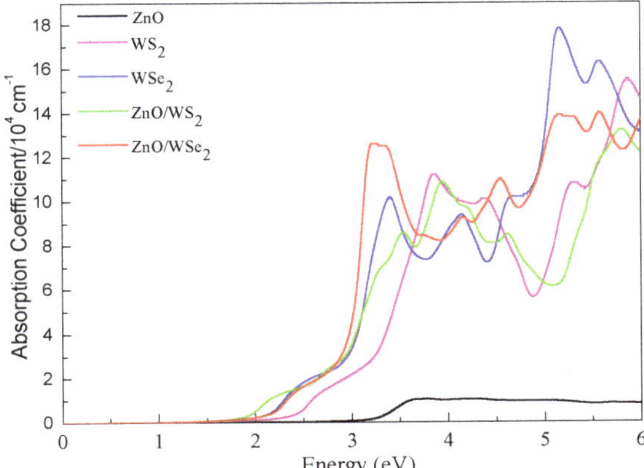

Figure 8. Calculated optical absorption curves of ZnO, WS_2, and WSe_2 monolayers and of ZnO/WS_2 and ZnO/WSe_2 heterostructures.

4. Conclusions

In summary, we perform extensive hybrid density functional calculation to examine the geometric, electronic, and optical properties as well as the band edge alignment of ZnO/WS_2 and ZnO/WSe_2 heterostructures and consider the possible effect caused by rotation angles and biaxial strains. ZnO/WS_2 and ZnO/WSe_2 with suitable rotation angles and strains are not difficult to prepare due to the negative interface adhesion energies. The bandgaps of these heterostructures are not obviously affected by the rotation angles, but they are by the strains. The band edge positions render ZnO/WSe_2 with different rotation angles and biaxial strains suitable for hydrogen generation but unfavorable for oxygen generation. ZnO/WS_2 with suitable rotation angles and strains have appropriate bandgaps for visible light absorbtion and proper band edge alignment for spontaneous water splitting. The charge transfer from the ZnO layer to the WS_2 layer will facilitate the separation of photogenerated carriers and improve the photocatalytic activity. These findings imply ZnO/WS_2 is a promising water-splitting photocatalyst.

Author Contributions: G.W. carried out the DFT calculations; Q.S. and S.D. analyzed the calculated results and produced the illustrations; G.W., S.X. and G.L. prepared the manuscript; D.L. and M.Z. designed and planned the research work, and guided G.W. et al. to complete the present work.

Acknowledgments: This work was supported by the National Natural Science Foundation of China under Grant No. 11504301, the Chongqing Key Laboratory of Additive Manufacturing Technology and Systems (CIGIT, CAS), and the Natural Science Foundation Project of Chongqing Science and Technology Commission under Grant No. cstc2016jcyjA1158.

Conflicts of Interest: The authors declare no conflict of interest.

References

1. Neto, A.H.C.; Guinea, F.; Peres, N.M.R.; Novoselov, K.S.; Geim, A.K. The electronic properties of graphene. *Rev. Mod. Phys.* **2009**, *81*, 109–162. [CrossRef]
2. Xiao, S.; Wang, T.; Liu, Y.; Xu, C.; Han, X.; Yan, X. Tunable light trapping and absorption enhancement with graphene ring arrays. *Phys. Chem. Chem. Phys.* **2016**, *18*, 26661–26669. [CrossRef] [PubMed]
3. Rahman, M.Z.; Kwong, C.W.; Davey, K.; Qiao, S.Z. 2D phosphorene as a water splitting photocatalyst: Fundamentals to applications. *Energy Environ. Sci.* **2016**, *9*, 709–728. [CrossRef]
4. Faccio, R.; Denis, P.A.; Pardo, H.; Goyenola, C.; Mombru, A.W. Mechanical properties of graphene nanoribbons. *J. Phys. Condens. Mat.* **2009**, *21*, 285304. [CrossRef] [PubMed]

5. Heersche, H.B.; Jarilloherrero, P.; Oostinga, J.B.; Vandersypen, L.M.K.; Morpurgo, A.F. Bipolar supercurrent in graphene. *Nature* **2007**, *446*, 56–59. [CrossRef] [PubMed]
6. Zhu, Y.; Murali, S.; Stoller, M.D.; Ganesh, K.J.; Cai, W.; Ferreira, P.J.; Pirkle, A.; Wallace, R.M.; Cychosz, K.A.; Thommes, M.; et al. Carbon-based supercapacitors produced by activation of graphene. *Science* **2011**, *332*, 1537–1541. [CrossRef] [PubMed]
7. Li, M.; Liu, Y.; Zhao, J.; Wang, X. Si clusters/defective graphene composites as Li-ion batteries anode materials: A density functional study. *Appl. Surf. Sci.* **2015**, *345*, 337–343. [CrossRef]
8. Zhang, P.; Hu, Z.; Wang, Y.; Qin, Y.; Sun, X.W.; Li, W.; Wang, J. Enhanced photovoltaic properties of dye-sensitized solar cell based on ultrathin 2D TiO_2 nanostructures. *Appl. Surf. Sci.* **2016**, *368*, 403–408. [CrossRef]
9. Putri, L.K.; Ong, W.; Chang, W.S.; Chai, S. Heteroatom doped graphene in photocatalysis: A review. *Appl. Surf. Sci.* **2015**, *358*, 2–14. [CrossRef]
10. Tusche, C.; Meyerheim, H.L.; Kirschner, J. Observation of depolarized ZnO(0001) monolayers: Formation of unreconstructed planar Sheets. *Phys. Rev. Lett.* **2007**, *99*, 026102. [CrossRef] [PubMed]
11. Claeyssens, F.; Freeman, C.L.; Allan, N.L.; Sun, Y.; Ashfold, M.N.R.; Harding, J.H. Growth of ZnO thin films: Experiment and theory. *J. Mater. Chem.* **2005**, *15*, 139–148. [CrossRef]
12. Das, R.; Rakshit, B.; Debnath, S.; Mahadevan, P. Microscopic model for the strain-driven direct to indirect band-gap transition in monolayer MoS_2 and ZnO. *Phys. Rev. B* **2014**, *89*, 106–112. [CrossRef]
13. Guo, H.; Zhao, Y.; Lu, N.; Kan, E.; Zeng, X.C.; Wu, X.; Yang, J. Tunable magnetism in a nonmetal-substituted ZnO monolayer: A first-principles study. *J. Phys. Chem. C* **2012**, *116*, 11336–11342. [CrossRef]
14. Qin, G.; Wang, X.; Zheng, J.; Kong, C.; Zeng, B. First-principles investigation of the electronic and magnetic properties of ZnO nanosheet with intrinsic defects. *Comp. Mater. Sci.* **2014**, *81*, 259–263. [CrossRef]
15. Schmidt, T.M.; Miwa, R.H.; Fazzio, A. Ferromagnetic coupling in a Co-doped graphenelike ZnO sheet. *Phys. Rev. B* **2010**, *81*, 195413. [CrossRef]
16. Fang, D.; Zhang, Y.; Zhang, S. Magnetism from 2p states in K-doped ZnO monolayer: A density functional study. *EPL* **2016**, *114*, 47012. [CrossRef]
17. Kaewmaraya, T.; De Sarkar, A.; Sa, B.; Sun, Z.; Ahuja, R. Strain-induced tunability of optical and photocatalytic properties of ZnO mono-layer nanosheet. *Comp. Mater. Sci.* **2014**, *91*, 38–42. [CrossRef]
18. Luo, X.; Wang, G.; Huang, Y.; Wang, B.; Yuan, H.; Chen, H. Bandgap engineering of the g-ZnO nanosheet via cationic-anionic passivated codoping for visible-light-driven photocatalysis. *J. Phys. Chem. C* **2017**, *121*, 18534–18543. [CrossRef]
19. Wang, G.; Chen, H.; Li, Y.; Kuang, A.; Yuan, H.; Wu, G. A hybrid density functional study on the visible light photocatalytic activity of (Mo,Cr)-N codoped $KNbO_3$. *Phys. Chem. Chem. Phys.* **2015**, *17*, 28743–28753. [CrossRef] [PubMed]
20. Wang, G.; Yuan, H.; Li, Y.; Kuang, A.; Chen, H. Enhancing visible light photocatalytic activity of $KNbO_3$ by N-F passivated co-doping for hydrogen generation by water splitting. *J. Mater. Sci.* **2017**, 1–12.
21. Moniz, S.J.A.; Shevlin, S.A.; Martin, D.J.; Guo, Z.; Tang, J. Visible-light driven heterojunction photocatalysts for water splitting-A critical review. *Energy Environ. Sci.* **2015**, *8*, 731–759. [CrossRef]
22. Low, J.; Cao, S.; Yu, J.; Wageh, S. Two-dimensional layered composite photocatalysts. *Chem. Commun.* **2014**, *50*, 10768–10777. [CrossRef] [PubMed]
23. Wang, H.; Li, X.; Yang, J. The g-C_3N_4/C_2N Nanocomposite: A g-C_3N_4-based water-splitting photocatalyst with enhanced energy efficiency. *ChemPhysChem* **2016**, *17*, 2100–2104. [CrossRef] [PubMed]
24. Liao, J.; Sa, B.; Zhou, J.; Ahuja, R.; Sun, Z. Design of high-efficiency visible-light photocatalysts for water splitting: MoS_2/AlN(GaN) heterostructures. *J. Phys. Chem. C* **2014**, *118*, 17594–17599. [CrossRef]
25. Wang, G.; Dang, S.; Zhang, P.; Xiao, S.; Wang, C.; Zhong, M. Hybrid density functional study on the photocatalytic properties of AlN/$MoSe_2$, AlN/WS_2, and AlN/WSe_2 heterostructures. *J. Phys. D Appl. Phys.* **2018**, *51*, 025109. [CrossRef]
26. Wang, G.; Yuan, H.; Chang, J.; Wang, B.; Kuang, A.; Chen, H. ZnO/MoX_2 (X = S, Se) composites used for visible light photocatalysis. *RSC Adv.* **2018**, *8*, 10828–10835. [CrossRef]
27. Kang, J.; Tongay, S.; Zhou, J.; Li, J.; Wu, J. Band offsets and heterostructures of two-dimensional semiconductors. *Appl. Phys. Lett.* **2013**, *102*, 012111. [CrossRef]
28. Kresse, G.; Furthmuller, J. Efficient iterative schemes for ab initio total-energy calculations using a plane-wave basis set. *Phys. Rev. B* **1996**, *54*, 11169–11186. [CrossRef]

29. Ernzerhof, M.; Scuseria, G.E. Assessment of the Perdew-Burke-Ernzerhof exchange-correlation functional. *J. Chem. Phys.* **1999**, *110*, 5029–5036. [CrossRef]
30. Perdew, J.P.; Burke, K.; Ernzerhof, M. Generalized Gradient Approximation made simple. *Phys. Rev. Lett.* **1996**, *77*, 3865–3868. [CrossRef] [PubMed]
31. Blochl, P.E. Projector augmented-wave method. *Phys. Rev. B* **1994**, *50*, 17953–17979. [CrossRef]
32. Grimme, S.; Antony, J.; Ehrlich, S.; Krieg, H. A consistent and accurate ab initio parametrization of density functional dispersion correction (DFT-D) for the 94 elements H-Pu. *J. Chem. Phys.* **2010**, *132*, 154104. [CrossRef] [PubMed]
33. Heyd, J.; Scuseria, G.E.; Ernzerhof, M. Hybrid functionals based on a screened Coulomb potential. *J. Chem. Phys.* **2003**, *118*, 8207–8215. [CrossRef]
34. Heyd, J.; Scuseria, G.E.; Ernzerhof, M. Erratum: "Hybrid functionals based on a screened Coulomb potential". *J. Chem. Phys.* **2006**, *124*, 219906. [CrossRef]
35. Saha, S.; Sinha, T.P.; Mookerjee, A. Electronic structure, chemical bonding, and optical properties of paraelectric $BaTiO_3$. *Phys. Rev. B* **2000**, *62*, 8828–8834. [CrossRef]
36. Lee, J.; Sorescu, D.C.; Deng, X. Tunable lattice constant and band gap of single- and few-layer ZnO. *J. Phys. Chem. Lett.* **2016**, *7*, 1335–1340. [CrossRef] [PubMed]
37. Bjorkman, T.; Gulans, A.; Krasheninnikov, A.V.; Nieminen, R.M. van der Waals bonding in layered compounds from advanced density-functional first-principles calculations. *Phys. Rev. Lett.* **2012**, *108*, 235502. [CrossRef] [PubMed]
38. Yu, W.; Zhu, Z.; Zhang, S.; Cai, X.; Wang, X.; Niu, C.; Zhang, W. Tunable electronic properties of GeSe/phosphorene heterostructure from first-principles study. *Appl. Phys. Lett.* **2016**, *109*, 103104. [CrossRef]
39. Li, S.; Wang, C.; Qiu, H. Single- and few-layer ZrS_2 as efficient photocatalysts for hydrogen production under visible light. *Int. J. Hydrogen Energy* **2015**, *40*, 15503–15509. [CrossRef]
40. Li, X.H.; Wang, B.J.; Cai, X.L.; Zhang, L.W.; Wang, G.D.; Ke, S.H. Tunable electronic properties of arsenene/GaS van der Waals heterostructures. *RSC Adv.* **2017**, *7*, 28393–28398. [CrossRef]

MDPI
St. Alban-Anlage 66
4052 Basel
Switzerland
Tel. +41 61 683 77 34
Fax +41 61 302 89 18
www.mdpi.com

Nanomaterials Editorial Office
E-mail: nanomaterials@mdpi.com
www.mdpi.com/journal/nanomaterials

www.ingramcontent.com/pod-product-compliance
Lightning Source LLC
LaVergne TN
LVHW070851170726
843515LV00005B/624

* 9 7 8 3 0 3 9 2 8 7 6 8 0 *